Community-Based Monitoring in the Arctic

Community-Based Monitoring in the Arctic

Finn Danielsen
Noor Johnson
Olivia Lee
Maryann Fidel
Lisbeth Iversen
Michael K. Poulsen

Hajo Eicken
Ania Albin
Simone G. Hansen
Peter L. Pulsifer
Peter Thorne
Martin Enghoff

Text © 2020 University of Alaska Press
Published by
University of Alaska Press
P.O. Box 756240
Fairbanks, AK 99775-6240

This book is produced with support from the EC H2020 project INTAROS (grant 727890), Danish Agency for Science and Higher Education (18/014779-13), and the National Science Foundation Belmont Forum Collaborative Research, Pan-Arctic Options, Holistic Integration for Arctic Coastal-Marine Sustainability (grant 1660520).

Preferred citation: Danielsen, F., N. Johnson, O. Lee, M. Fidel, L. Iversen, M. K. Poulsen, H. Eicken, A. Albin, S. G. Hansen, P. L. Pulsifer, P. Thorne, M. Enghoff. 2020. *Community-Based Monitoring in the Arctic*. University of Alaska Press.

Library of Congress Cataloging-in-Publication Data

Names: Danielsen, Finn (Dr.scient., Ecologist), author.
Title: Community-based monitoring in the Arctic / Finn Danielsen, Noor Johnson, Olivia Lee, Maryann Fidel, Lisbeth Iversen, Michael K. Poulsen, Hajo Eicken, Ania Albin, Simone G. Hansen, Peter L. Pulsifer, Peter Thorne, and Martin Enghoff.
Description: Fairbanks : University of Alaska Press, 2020. | Includes bibliographical references and index. | Summary: "Arctic people live in and observe the Arctic environment year-round. Intimate knowledge of the environment and environmental changes is fundamental to their survival. They frequently depend on natural resources for their livelihood. This book presents a review of the capabilities, good practices, opportunities and barriers of community-based environmental monitoring programs in the Arctic, with a focus on decision-making for resource management"— Provided by publisher.
Identifiers: LCCN 2020001551 (print) | LCCN 2020001552 (ebook) | ISBN 9781602234284 (paperback) | ISBN 9781602234291 (ebook)
Subjects: LCSH: Environmental monitoring—Arctic regions. | Natural resources—Co-management—Arctic regions.
Classification: LCC QH541.15.M64 D36 2020 (print) | LCC QH541.15.M64 (ebook) | DDC 363.7/06309113—dc23
LC record available at https://lccn.loc.gov/2020001551
LC ebook record available at https://lccn.loc.gov/2020001552

Cover and interior design by Paula Elmes.
Cover image: Fisherman hauling in a Greenland halibut (*Reinhardtius hippoglossoides*) in Disko Bay, Greenland. Photo by Finn Danielsen.

Educational use:
This book (in part or in its entirety) and products available from the UArctic Thematic Network on Collaborative Resource Management (www.uarctic.org/organization/thematic-networks/collaborative-resource-management/) can be used freely as teaching materials and for other educational purposes. The only condition of such use is acknowledgment of the source of the material according to the recommended citation. In case of questions regarding educational use, please contact the Thematic Network. Any views expressed in this peer-reviewed book are the responsibility of the authors of the book and do not necessarily reflect the views of the contributing institutions or funding institutions.

Contents

Acknowledgments, vii

Summary, ix

1 Introduction 1
1.1 Methods, 3

2 Capabilities 9
2.1 Introduction, 9
2.2 Characteristics of Arctic CBM Programs, 10
2.3 Coverage of Arctic CBM Programs, 13
2.4 The Framework and Format of Knowledge Products from the Arctic CBM Programs, 19
2.5 Linkages to International Agreements and UN Sustainable Development Goals, 21
2.6 Summary: Capabilities, 24

3 Good Practice 27
3.1 Good Practice: Establishing CBM Programs, 28
3.2 Good Practice: Implementing CBM Programs, 32
3.3 Good Practice: Sustaining CBM Programs, 36
3.4 Good Practice: Obtaining Impacts Through CBM, 38
3.5 Good Practice: Connecting and Cross-Weaving with Other Approaches, 41
3.6 Good Practice: Ensuring the Quality of Knowledge Products, 46
3.7 Good Practice: Addressing the Rights of Indigenous and Local Communities, 53
3.8 Summary: Good Practice, 57

4 Challenges and How to Address Them 59
4.1 Challenge: Limited Ability or Interest of Management Agencies to Access, Understand, and Act on CBM-Derived Guidance, 60
4.2 Challenge: Insufficient Linkages Between CBM Programs and the Priorities of Northern Communities, 64
4.3 Challenge: Sustaining Community Members' Long-Term Commitment to CBM Efforts, 66
4.4 Challenge: Lack of Compatibility Between Data Formats of Scientist-Executed Monitoring and CBM Programs, 68
4.5 Challenge: Intellectual Property Rights, Respect and Reciprocity, and Free, Prior, and Informed Consent (FPIC), 70
4.6 Challenge: Organizational and Support Structures for CBM Programs, 73
4.7 Summary: Challenges and How to Address Them, 75

5 Moving Forward with CBM in the Arctic 77

Appendix A. Arctic CBM Programs, 87

Appendix B. CBM Practitioner Questionnaire, 91

References, 99

Index, 109

Acknowledgments

The authors are very grateful for the contribution of a number of highly experienced community members, practitioners and scholars engaged in community-based monitoring:

Achim Drebs, Bjarne (Ababsi) Lyberth, Donna Hauser, Dögg Matthíasdóttir, Eduard Zdor, Else Grete Broderstad, Eyðfinn Magnussen, Göran Ericsson, Heather Ashthorn, Jan Sørensen, Jason Akearok, Jessica Aquino, José Gérin-Lajoie, Katie Spellman, Karl Zinglersen, Leif Jougda, Lene K. Holm, Lisa Sheffield Guy, Marianne Rasmussen, Mark Nuttall, Michael Birlea, Michael Brubaker, Nette Levermann, Nikita Vronski, Pâviârak Jakobsen, Per Ole Frederiksen, Pekka Helle, Rita Johansen, Robert T. Barrett, Rodion Sulyandziga, Tero Mustonen, Vyacheslav (Slava) Shadrin, Weronika Axelsson Linkowski, Zoya Martin.

External review of the text was undertaken by Kirsten M. Silvius and Henry P. Huntington.

This document was produced with support from the EC H2020 project INTAROS (grant 727890), Danish Agency for Science and Higher Education (18/014779-13), the National Science Foundation Belmont Forum Collaborative Research, Pan-Arctic Options, Holistic Integration for Arctic Coastal-Marine Sustainability (grant 1660520), and by financial contributions made by the institutions of all of the participating authors. Claire Armstrong, Hiroyuki Enomoto, Roberta Pirazzini, Hanne Sagen, and Stein Sandven provided insightful guidance.

Graphics and layout by Paula Elmes, ImageCraft Publications & Design.

■ **Summary**

Arctic people live in and observe the Arctic environment year-round. Intimate knowledge of the environment is fundamental to their survival. They frequently depend on natural resources for their livelihood. This book presents a review of the capabilities, good practices, opportunities, and barriers of community-based environmental monitoring programs in the Arctic, with a focus on decision-making for resource management.

The review builds upon previous work that contributed to the Sustaining Arctic Observing Networks (SAON, Task #9 n.d.; Johnson et al. 2016). We first identified 170 community-based monitoring programs in the Arctic from the peer-reviewed literature and from searching the internet. We then chose 30 programs that reflected the widest possible set of situations and issues for a more in-depth analysis. We reviewed the scientific literature and discussed experiences at workshops with practitioners and community members engaged in monitoring programs in Nuuk, Fairbanks, Québec, Longyearbyen, and the Russian districts of Komi Izhma, Zhigansk, and Olenek between 2017 and 2019. The key findings are summarized below.

Capabilities

There is a long history of involving community members of all ages in monitoring the Arctic. Programs involve organizations such as community groups, all levels of government, universities, schools, and the private sector. Programs monitor biological attributes, abiotic phenomena, and sociocultural attributes, often within the same framework. The observing domains that receive the greatest attention are "land and cryosphere" and "ocean and sea ice." By their nature, community-based monitoring programs tend to focus on those issues of greatest concern to local stakeholders; thus, outcomes from such observing programs have considerable potential to influence on-the-ground management activities.

The programs complement scientist-executed monitoring by using different methodologies and engaging the experience of Indigenous knowledge holders and other long-term residents who have significant knowledge of the environment. Many of the community monitoring programs also complement existing research observations by providing an increase in sample size or density, area, and time. Most Arctic community-based monitoring programs make observations between 61° N and 70° N. They cover all the Arctic biomes with the exception of ice desert. Data are typically collected throughout the year. The majority of the programs involve Indigenous knowledge. Some programs inform decisions at local, regional, and national levels, particularly when there are connections to agencies that respect the rights of resource-dependent communities in shaping decision-making. These programs often provide insight into processes and changes not captured in government agency or research-driven monitoring programs. Thus, community-based monitoring programs can contribute to better-informed decisions or better-documented processes within the key economic sectors in the Arctic:

- Fisheries
- Forestry
- Herding
- Hunting
- Mineral and hydrocarbon extraction
- Shipping
- Tourism

Methods originating from both the natural and social sciences are often used, and methods that draw on Indigenous knowledge are also increasingly used in program design. New technologies enable the programs to collect data and communicate findings with greater certainty than ever before. Some programs have made their data publicly available, but few have links to data discovery portals or global repositories. Most of the CBM program organizers reported that the programs have helped communities advance their own social and political goals. Community monitoring programs have the potential to contribute to achieving the objectives of 10 international environmental agreements that have particular relevance to the Arctic. They could also contribute to achieving 16 of the 17 United Nations Sustainable Development Goals.

Good Practices

Arctic community-based monitoring programs are diverse, with many successful approaches. By providing actionable information to management authorities and community members, the programs inform many kinds of decisions. Web-based knowledge management platforms are increasingly used for data storage and communication. Credible knowledge products are obtained in many ways, including through careful planning, thorough guidance of the participants, and validation of data using different approaches. Many programs follow the principles of "free, prior, and informed consent" and contribute, directly or indirectly, to protecting the rights of the Indigenous and local communities. Co-design of programs and co-production of knowledge can help ensure the relevance and utility of monitoring data.

Opportunities and Barriers

Community-based monitoring programs have strong potential for linking environmental monitoring to awareness-raising, capacity-building, and enhanced decision-making at all levels of resource management. Knowledge developed over time by communities of practice is valuable for the management of natural resources. Moreover, community-based monitoring programs could be used to develop new hypotheses, provide data that could be used to fill gaps in climate modeling, and contribute to research in areas such as risk management, safety, and food and water security. Community-based monitoring contributes to fulfill the rights of citizens to take part in decisions related to their local and regional areas, and allows them to participate in knowledge production efforts used to develop and safeguard their environment.

One barrier to maximizing the potential of community-based monitoring for decision-making has been the perception that information from local people is subjective and anecdotal. Today, a growing body of literature demonstrates that where Indigenous and local knowledge has been systematically gathered, the data collected by community members are comparable to those arising from professional scientists. Another barrier is that management authorities are sometimes slow at operationalizing or acting upon community observations in their decision-making. Lack of recognition and trust remains a substantial barrier for successful collaboration. Regardless of the barriers described, involving people who face the daily consequences of environmental change in the Arctic is important for providing relevant information that can be used for adapting natural resource management to the local realities in a rapidly changing Arctic.

1 Introduction

At a workshop in Fairbanks in May 2017, community members, practitioners, and scientists from Alaska summarized the value and role of community-based monitoring (CBM) in the Arctic. They used the following words:

> The challenges of our time call for greater, more effective collaboration. Environmental change is occurring rapidly. There is an urgency to the situation, a climate crisis. This makes the community-based monitoring and documentation of Indigenous and local knowledge more important than ever.
>
> Arctic Indigenous Peoples find themselves not only at the "front-lines" of climate change impacts, but are at the front-lines of creating hypotheses about change and adaptation. It is human nature when seeing a change to think about why that change is happening. . . . Baseline data is often lacking. Indigenous and local knowledge can sometimes fill the gap.
>
> Indigenous peoples are often the first to see change since they are traveling, and harvesting on the landscape. They can also provide historical context due to long-term intimacy with the environment, and importantly answer why the change matters. They know when it impacts community and culture. . . . It's not just academic curiosity to understand a phenomenon. People are already acting and adapting to changing conditions. . . .

> The gap between information and action needs to be shortened. Information is needed to make choices. Information needs to be in the hands of people who are adapting. Community-based monitoring can shorten the gap between research and action, by empowering Indigenous peoples to collect data to address decision making needs. (Fidel et al. 2017)

The Fairbanks workshop was one of a series of workshops on community-based environmental monitoring in the Arctic funded by the European Union Horizon 2020 Program as part of the Integrated Arctic Observation System Project (INTAROS). The other workshops were held in Nuuk; the Russian districts Komi Izhma, Zhigansk, and Olenek; Québec City, Canada; and Longyearbyen, Norway (Enghoff et al. 2019; Johnson et al. 2018; Poulsen et al. 2019; see Table 1.1).

This project aims to extend and improve existing and evolving observing systems that encompass land, air, and sea in the Arctic (INTAROS 2020). One of the project components focuses on enhancing community-based observing in the Arctic. Key activities include knowledge exchange workshops, exploring opportunities to interweave existing CBM programs in the Arctic with scientists' monitoring efforts, and piloting new tools in Greenland and Svalbard to support decision-making and capacity-building.

In many areas of the Arctic, civil society organizations, government agencies, or researchers have established CBM programs. The programs use Indigenous and local knowledge and observations, and build upon existing community-based approaches to observing the environment. A recent review of the Sustaining Arctic Observing Networks (SAON) analyzed a sample of CBM programs in the Arctic (SAON, Task #9). One of the objectives of INTAROS is to survey and analyze existing CBM programs in the Arctic and identify capabilities, good practices, and challenges while building on the review that contributed to SAON (Johnson et al. 2016).

This book presents the results of the survey of Arctic CBM programs and offers analysis based on survey results as well as observations contributed by practitioners at the workshops listed above. First, we describe the characteristics of the programs, their coverage, the framework and format of their knowledge products, and the potential linkages to sustainable development goals (Chapter 2). Second, we summarize good practices in terms of implementing and sustaining the programs, obtaining impacts, connecting with other approaches, ensuring the quality of knowledge products, and addressing the rights of Indigenous and local communities (Chapter 3). We conclude with a

Fishers and hunters in Attu, Greenland, discuss their observations and trends in resources, and propose management interventions to the municipal authorities. For example, in 2017, they proposed that game officers should step up their effort at Naternaq (Lersletten) "to ensure muskox is only harvested in accordance with the legislation." Subsequently, the government established a moratorium on muskox (*Ovibos moschatus*) hunting in this area until population surveys are undertaken. Sometimes, proposals from CBM programs benefit the people who initiate them, while at other times community members suggest restrictions on their own take of resources. Credit: Michael K. Poulsen and PISUNA

discussion of the challenges that Arctic CBM programs face, and how to address them (Chapter 4). Finally, we discuss how to move forward with CBM in the Arctic (Chapter 5).

1.1 Methods

The book is based on a self-reporting survey of CBM programs (Chapter 2), a review of the scientific literature, and workshops with CBM program practitioners and community members engaged in CBM programs (Chapters 3 and 4). The survey, the literature review, and the workshops build upon the previous work of SAON (Johnson et al. 2016; see Figure 1.1).

The previous work undertaken as a contribution to SAON was based on workshops held in 2013–2014 (Johnson et al. 2016). In the present study, we take a renewed look at the landscape of Arctic CBM programs. With fresh eyes, we reexamined the fundamental aspects of the programs in much greater detail for a smaller set of CBM programs. This approach allowed us to investigate topics such as the attributes covered by CBM programs, the field tools used, the role of Indigenous knowledge, and how to obtain results, connect with other approaches, address the rights of Indigenous communities, and address a range of challenges for sustaining these programs. As a result, we were able to update and expand our analysis and cover several aspects of the Arctic CBM programs that have received only limited attention in the past. We paid particular attention to developments over the past three to six years where the Arctic has undergone rapid sociocultural and environmental changes. We focused on the role of CBM programs for informing decision-making in resource management.

For the self-reporting survey of CBM programs, we first identified 170 CBM programs in the Arctic from the peer-reviewed literature and from searching the internet. We chose 45 programs for a more in-depth analysis to reflect the widest possible set of situations and issues. This included criteria for wide geographical coverage and a breadth of attributes being monitored. We prepared a multiple-choice questionnaire with 35 questions directed at the organizers of each program. There was one respondent for each CBM program. Thirty out of the 45 CBM programs targeted completed the survey (Annex A), resulting in a 67% survey response rate. Sixteen of the 30 CBM programs were assessed for the first time in this study. Therefore, the study results include an assessment of a large number of programs that were not assessed in the earlier survey by Johnson et al. (2016). CBM practitioners were asked general questions relevant to all Arctic monitoring systems, as well as questions of particular relevance to CBM programs. The general questions were about the respondent, the characteristics of the attributes observed, the sustainability of the monitoring, and the use and management of the data. The CBM-specific survey questions addressed key features of each CBM program (aims, community engagement, data derived from the program, linkages to natural resource governance and decision-making, and key challenges; see Annex B). We used the results to describe the characteristics and coverage of the CBM programs, as well as to identify the framework and format of knowledge products that came from the CBM programs.

For each CBM program, we then assessed: (1) its ability to contribute, or probably contribute, to better-informed decisions and better-informed processes in key economic sectors in the Arctic region; (2) which CBM programs

Figure 1.1. Location of community-based monitoring programs identified for the Sustaining Arctic Observing Networks (adapted from Johnson et al. 2016; www.arcticcbm.org).

could, or probably could, contribute to achieving the objectives of 10 multilateral agreements in the Arctic; and (3) which of the 17 UN Sustainable Development Goals the CBM programs could contribute to achieving. All CBM programs were assessed by one coauthor; in the few cases when there was doubt in the evaluation of relevance to policy or decision-making, consensus was reached through discussion with other coauthors.

The workshops with CBM program practitioners and community members were held in Alaska, Canada, Greenland, Svalbard, and Russia from 2016 to 2019 (Table 1.1). Workshop participants included representatives from 14 CBM programs that also participated in the survey (Annex A), as well as some that did not. As such, some of the examples included in this book were not included in our analysis of survey data. We have noted those that were not part of the survey where they appear in the text in Chapter 2.

Topics discussed during the experience exchange workshops varied as organizers considered participant interests. In general, topics included how to sustain CBM activity, who uses the information generated and how, whether there was interest in sharing the information with others beyond current users of the CBM program, and the barriers and opportunities that exist for doing so. Separate proceedings have been prepared for most of the workshops (Table 1.1).

Table 1.1. Experience exchange workshops engaging community-based monitoring practitioners and community members (2017–2019)

Workshop	Dates	Host	Proceedings
Nuuk, Greenland	December 6–8, 2016	NORDECO, Piniakkanik Sumiiffinni Nalunaarsuineq (PISUNA), and NUNAVIS	Unpublished report
Fairbanks, Alaska, United States	May 10, 2017	International Arctic Research Center, University of Alaska Fairbanks, Yukon River Inter-Tribal Watershed Council, ELOKA, and NORDECO	Fidel et al. 2017
Québec City, Québec, Canada	December 11–12, 2017	ELOKA, Yukon River Inter-Tribal Watershed Council, NORDECO, Nansen Environmental and Remote Sensing Center, University of Alaska Fairbanks	Johnson et al. 2018
Komi Izhma, Zhigansk and Olenek Districts, Russia	September 2017, September 2018, April 2019	Centre for Support to Indigenous Peoples of the North, the Republic Indigenous Peoples Organization of Sakha Republic, NORDECO	Enghoff et al. 2019
Longyearbyen, Svalbard, Norway	March 7–8, 2019	Nansen Environmental and Remote Sensing Center and NORDECO	Poulsen et al. 2019

Local observer examining the water under the sea ice in Kotzebue, Alaska.
Credit: Donna Hauser

We intentionally did not predefine CBM but instead adopted an inclusive approach that encompassed programs with different levels of community involvement as well as Indigenous and local knowledge documentation projects with relevance to long-term observing (that is, those that document environmental attributes that may change over time and would be able to serve as a baseline for observing change). Of the 30 CBM programs included in the survey, 17% were traditional knowledge documentation projects that did not include repeated data collection over time. Further discussion of the definitions of CBM is available in Johnson et al. (2016) and summarized in Box 1.1. We used the Inuit Circumpolar Council's definition of Indigenous knowledge: "a systematic way of thinking applied to phenomena across biological, physical, cultural and spiritual systems. It includes insights based on evidence acquired through direct and long-term experiences and extensive and multigenerational observations, lessons and skills" (Inuit Circumpolar Council n.d.).

BOX 1.1

Definitions of Community-Based Monitoring

CBM PROGRAMS ARE DISTINGUISHED from scientist-executed monitoring programs by the involvement of community members in one or more steps of the monitoring process. Definitions of community-based monitoring can encompass a range of local community or stakeholder involvement. On the broader spectrum, this could be defined as "a process where concerned citizens, government agencies, industry, academia, community groups, and local institutions collaborate to monitor, track, and respond to issues of common community concern" (EMAN 2003:4). An emphasis on community interests in monitoring objectives was also provided in Danielsen et al.'s definition (2014d:15) of CBM as "monitoring . . . undertaken by local stakeholders using their own resources and in relation to aims and objectives that make sense to them." Arctic CBM programs that involve Indigenous communities have an added distinction of often including traditional and Indigenous knowledge, although this was not a requirement for inclusion in this survey of Arctic CBM programs.

The general understanding of CBM programs included in this survey follows the definition of Johnson et al. (2015:29), where CBM is "a process of routinely observing environmental or social phenomena, or both, that is led and undertaken by community members and can involve external collaboration and support of visiting researchers and government agencies." This definition did not require programs to have been explicitly co-developed with community collaboration, differing from some definitions of Arctic community-based observing networks (Griffith, Alessa, and Kliskey 2018).

■ 2 Capabilities

2.1 Introduction

The international observing community has invested in efforts to establish and maintain an inventory of Arctic CBM programs, but in the effort to capture as many programs as possible, there has been less emphasis on describing more nuanced characteristics of CBM efforts. In this chapter, we describe in more detail some of the characteristics of Arctic CBM programs and assess their capabilities in terms of coverage, framework, and format of knowledge products. We discuss the potential linkages to decision-making at various scales, as well as the relevance to broader goals in international agreements and UN Sustainable Development Goals in the Arctic. Putting Arctic CBM outcomes in context with international agreements may also help leverage efforts to obtain external funding support for CBM while providing recognition of the key role that Arctic CBM participants play in remote locations, which are often data-sparse.

The chapter is based on a self-reporting survey of CBM programs, described in Chapter 1. Organizers of 30 CBM programs (Annex A) completed a multiple-choice questionnaire with 35 questions (Annex B). The survey was part of a broader INTAROS effort to survey observing efforts in the Arctic, including satellite remote-sensing data used to monitor environmental change in the Arctic. We adjusted all the survey questions to use terminology consistent with that used by CBM practitioners.

While steps were taken to ensure objectivity, some aspects of this work remain subjective or impacted by sampling errors. Thus, only successful CBM programs with active engagement were likely to respond to the survey. When identifying CBM programs we also relied on the peer-reviewed literature, which may underestimate small-scale local and volunteer-based efforts, likely biasing our findings toward larger, well-funded programs. Overall, we consider the magnitude of our estimates and their relative proportions acceptable for the purposes of this book.

2.2 Characteristics of Arctic CBM Programs

History

There is a long history of CBM programs in the Arctic. Although most of the programs that we assessed were established within the last 15 years (70% of 30 respondents; Figure 2.1), there are CBM programs that are far older. For instance, in 1932, the Federation of Icelandic River Owners began monitoring salmon (*Salmo salar*) and trout (*Salvelinus alpinus*) fished in rivers in Iceland. In 1919, the Finnish Meteorological Institute initiated monitoring of the changing year-to-year snow coverage as a CBM program. To our knowledge, the longest-running Arctic CBM program is held by community members in the Faroe Islands, in the northeast Atlantic Ocean, who began tracking and reporting their catch of pilot whales (*Globicephala melas*) to the government in 1584. Our survey results include some of the oldest and longest-running Arctic CBM programs, including three programs that began more than 60 years ago.

Figure 2.1. Age of the Arctic community-based monitoring programs (*n* = 30).

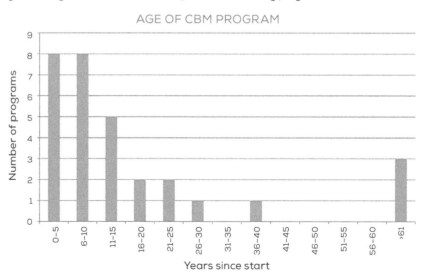

Community Involvement

The central feature that distinguishes CBM programs from scientist-executed monitoring programs is the involvement of community members in one or more steps of the monitoring process (see Box 1.1 and "Methods" in Chapter 1). Most of the CBM programs we assessed had fewer than 100 community members involved (67%), although some had between 100 and 1,000 members (20%), or more than 1,000 members (13%).

In most CBM programs, community members themselves had proposed their involvement (77%), but sometimes they were appointed by community leaders or other bodies to participate because of their background (40%). Occasionally, community members participated because they were obliged to if they wanted to fish or hunt in an area. Several programs included a mix of recruitment approaches; hence, the total proportion for community involvement types exceeds 100%.

Why community members were interested in being involved in the CBM programs: Our assessment, which is based on one respondent from each CBM program, suggests that there are often several reasons for community members to participate in CBM programs: Community members mostly wanted to have their voices heard and wanted to protect their rights over land and resources (60%); they also wanted to sustain wildlife health and abundance

Community members in Venetie, Alaska, concerned about the quality of the river water have begun monitoring it. Credit: Maryann Fidel

(57%). Less common reasons were social engagement (43%), personal benefits (36%), leisure interest (21%), or simply that it was mandatory to participate (11%). In a few cases, community members participated because it was an opportunity for learning and contributing to a rebirth of traditional practices and knowledge.

Although community members were motivated to engage in the CBM program, participation took time away from fishing and other livelihood activities. Roughly half the CBM programs provided compensation to community members for their time (52% of 29 respondents). Since there were no follow-up surveys, we were unable to determine whether availability of compensation had any effect on the long-term commitment of community members to participate in the CBM programs.

Gender representation: Overall, men were overrepresented in the programs. Half of 24 programs responding had less than 25% women among their participants. A potential source of this disparity may include the prevalence of CBM programs that monitor biological resources, as mostly men are engaged with hunting and fishing. This issue has been highlighted by others who recognized the general engagement of Inuit men and hunters with programs aimed at monitoring Arctic environmental change (Dowsley et al. 2010). However, there were also examples of programs that mostly involved women—for instance, Nordland Ærfugl in the Vega islands off Norway has a monitoring effort where landowners exploit eider (*Somateria mollissima*) down and register breeding eider ducks (eiderducks.no). More than 75% of participants are women. However, from our dataset we were unable to tell whether the roles and tasks within the individual CBM programs differed by gender.

Age: Most programs (66%) involved a mix of youth (19–26 years), adults (27–60 years), and seniors (> 60 years). Some also involved children (32%). CBM programs that provide resources for educators can be important vehicles for encouraging children's participation. For example, the Winterberry project in Alaska provides lesson plans that teachers can download. Other considerations for engaging with Indigenous knowledge holders across age groups are described in section 3.7.

Capacity Enhancement

We questioned whether the Arctic CBM programs contribute positively or negatively to the local community. We found that CBM programs helped communities advance their own social and political goals, and supported development of pride and self-esteem on the part of the individuals involved.

The Arctic CBM programs particularly contributed to three dimensions of capacity enhancement:

1. Social enhancement (67% of 30 respondents; e.g., education, and improvement of local organizations involved in managing natural resources).
2. Political enhancement (60%; e.g., participation in natural resource decision-making, leadership development, and increased local governance over natural resources).
3. Cognitive enhancement (60%; e.g., development of pride and self-esteem in natural resource management).

A number of the CBM programs also contributed to the economic enhancement of the local community (43%; e.g., financial resources and control of subsistence resources). No CBM programs reported contributing negatively to community capacity, although the organizers of a few programs did not know if the monitoring had enhanced the capacity of the local community (7%). It is worth noting that our findings are based on the self-reporting of a single-point contact from each CBM program (usually one of the organizers). Hence, the results provide a preliminary understanding of the benefits accrued to the local communities from successful Arctic CBM programs with respect to capacity enhancement. Further work is needed to verify and substantiate these findings in more detail.

2.3 Coverage of Arctic CBM Programs

Attributes

Arctic CBM programs monitored biological (93%), abiotic (50%), and sociocultural attributes (43%) (Figure 2.2). Many of the biological attributes monitored (e.g., mammals, birds, fish, shellfish, insects, plants, fungi) relate to goods and services provided by the natural environment or ecosystems. Abiotic phenomena included sea ice, water, snow, weather, wind, air quality, contaminants, ocean currents, and infrastructure development. Sociocultural attributes included Indigenous knowledge (about multiple topics) as well as language, human health, and wellness. Several of the programs monitored biological, abiotic, and sociocultural attributes at the same time (33%). More detailed examples of attributes monitored in the CBM programs are given in Chapter 3.

Observing Domains

We assessed the CBM programs' coverage of the three main observing domains: atmosphere, ocean/sea ice, and land/terrestrial cryosphere. We found

Figure 2.2. Proportion of Arctic community-based monitoring programs that monitor biological attributes, abiotic phenomena, and sociocultural attributes (*n* = 30 programs).

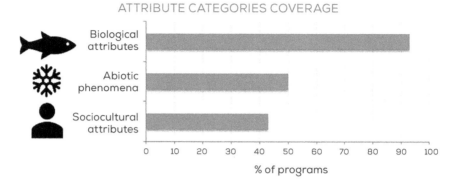

that 83% of the CBM programs covered one domain, and 17% covered two domains. Most CBM programs covered the land domain, including the terrestrial cryosphere (80%), 33% covered the ocean and sea ice, and 27% covered the atmosphere. Despite the low coverage of atmospheric attributes in the surveyed programs, the atmosphere is the focus of citizen science networks that collect data from citizens' weather stations, some of which are found in the Arctic (wow.metoffice.gov.uk/; www.cocorahs.org/). These networks were not part of our survey. There are also initiatives where students are involved in the rescue of historical weather data, for example, from the early whaling expeditions that provide historical observations of the sea ice edge (www.oldweather.org). This finding emphasizes that domain-specific interests in CBM coverage may require further analysis of a broader range of programs (for example, the project involving data rescue from historic whaling records was not included as one of the surveyed programs).

Geographic Distribution

There are CBM programs in all countries of the Arctic (Johnson et al. 2016), and the Atlas of Community-Based Monitoring and Indigenous Knowledge in a Changing Arctic aims to keep track of them (www.arcticcbm.org). The majority of programs in our sample made observations between the latitudes 61° and 70° N (> 50%; Figures 2.3–2.4). Some made observations farther north of 70° N. For example, Piniakkanik Sumiiffinni Nalunaarsuineq (PISUNA) include observations in the fishing and hunting areas of Qaanaaq, Greenland, at 77° N. Likewise, guides and guests on expedition cruises make environmental observations north of Svalbard, off Norway (76–81° N; Poulsen et al. 2019). A notably reduced number of observations occurs in the central part of the

Figure 2.3. Location of the Arctic community-based monitoring programs in our sample (*n* = 30). Map by Simone G. Hansen.

Figure 2.4. Latitudes where the Arctic community-based monitoring programs in our sample made environmental observations (*n* = 30).

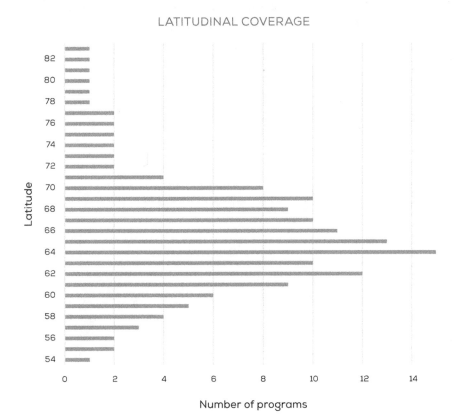

Arctic Ocean, which is rarely visited by community members. The CBM programs in our sample provide few data or information from this area (> 81° N).

Biomes

Programs in our sample covered taiga (boreal forest), tundra, freshwater, coastal areas, and marine areas (Figure 2.5). Ice desert is a characteristic of the interior of Greenland, Iceland, Svalbard, and Novaya Zemlya off the north coast of Siberia. Community members, however, rarely visit the ice deserts, and minimal CBM-derived information is collected from this biome.

Temporal Coverage

Community members generally undertook environmental observations throughout the year (Figure 2.6). One in three of the programs in our sample had no distinct data collection period; data are collected year-round (37%). Some had distinct time periods for data collection, and of these, there were peaks in the number of programs that collected data between the months of May/June and September/October. The ability to collect data year-round is an advantage of CBM programs, allowing more complete time series of records than may be obtained from scientist-executed monitoring alone (Mahoney and Gearheard 2008).

The Role of Indigenous Knowledge

The majority of programs in our sample (53%) engaged Indigenous knowledge in some capacity. Of those that did not (30%), the efforts typically involved contributory citizen science initiatives (Bonney et al. 2014) that engaged volunteers in collecting data for scientific research and monitoring. These citizen science initiatives were often not based specifically in the Arctic but were either active over the entire country or in regions outside of the Arctic (Johnson et al. 2016). (See further discussion of Indigenous and local knowledge in sections 3.7 and 4.5.)

Linkages to Economic Sectors

By their nature, CBM programs tend to focus on issues of greatest concern to local stakeholders. We therefore assessed the programs' linkages to key economic sectors in the region, and we found that they contributed to better-informed decisions or better-documented processes in several economic sectors in the Arctic (Table 2.2). The programs had the capacity to contribute to sectors of hunting/herding (60%), forestry (47%), fisheries (40%), shipping (37%), tourism (37%), and mineral and hydrocarbon extraction (20%).

Figure 2.5. Biomes covered by the Arctic community-based monitoring programs (n = 30).

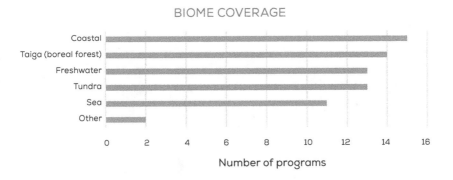

Figure 2.6. Time of year when the Arctic community-based monitoring programs undertook environmental observations (n = 30).

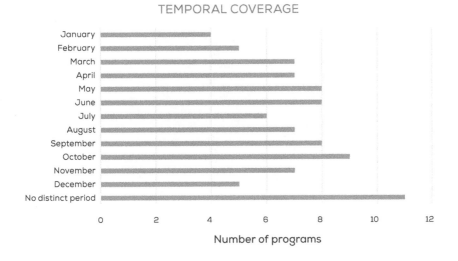

Across the Arctic countries, the importance of each economic sector to the national, regional, and local economy varies, with substantial variation within each country. For example, mineral and hydrocarbon extraction is important to the local economy on the North Slope of Alaska but is not significant for the local economy in the Bering Sea region in Alaska. In comparison, commercial fisheries are the dominant economic sector in the Bering Sea region but have not yet been developed on the North Slope. Further work is needed to understand how the characteristics of the economy influence local or regional government decisions to invest in and sustain CBM programs.

Table 2.2. The contribution of Arctic community-based monitoring programs to key economic sectors in the Arctic (blue = contributed; orange = probably contributed; $n = 30$; Annex A).

Program	Fisheries	Forestry	Herding and hunting	Mineral and hydrocarbon extraction	Shipping	Tourism
A-OK			blue		orange	
Arctic and Earth SIGNs	blue	orange			blue	
Arctic Borderlands	blue		blue			
Bird Phenology						
BuSK			blue			
CSIPN	blue	orange	blue			
Evenk & Izhma Peoples			blue			
Farmers and Herders			orange			
Fávllis	blue		blue		blue	
FMI Snow Depth		orange				
Fuglavernd						
George River	blue				orange	orange
Hares			blue			
ION-Yukon River	orange					
LEO	orange	orange	orange	orange	orange	orange
Marion Watershed	blue			orange		
Nordland Ærfugl						
Oral History	blue		orange			
Pilot Whale			blue		orange	
Piniarneq			blue			
PISUNA	blue		blue			
Renbruksplan		orange	blue			
River Owners Iceland	blue					
SIWO					blue	
Seal Monitoring					orange	orange
Walrus Haulout			blue		orange	
Wildlife Triangles		orange	blue			
Wild North					orange	orange
WinterBerry		orange				
Älgdata			blue			

2.4 The Framework and Format of Knowledge Products from the Arctic CBM Programs

Field Tools

Arctic CBM programs collect data in a variety of ways (Johnson et al. 2016), and the methods used influence the format of the knowledge products. Broadly speaking, these can be clustered into two approaches: one involving repeated measurements, and one emphasizing more qualitative measures (further discussed in section 3.6). The first approach involves collecting physical or biological samples, measurements, and observations. Most of the surveyed CBM programs used this approach (83%). The other more qualitative approach could be based on perceptions from focus group discussions, workshops, and use of oral history telling and interviews (40%). A total of 27% of the CBM programs surveyed combined these two approaches, using both sample- and perception-based approaches.

Time from Data Collection to Availability of the Results to the Users

CBM programs have the potential to quickly lead to decision-making (Danielsen et al. 2010; Danielsen et al. in review). Delays in data availability can be particularly crucial for decision-making processes in rapidly changing environments such as the Arctic. We assessed how long the Arctic CBM programs took from data collection until the results were available to the users. Of 25 respondents, 44% reported it took less than a month from data collection until the data were available to the users. For another 32%, it took between 6 and 12 months.

Data Validation

Data validation approaches were built into the field data sampling strategies of 50% of the CBM monitoring programs. It was most common for the CBM programs to triangulate their data across community members or across methods (both approaches 27%). Less frequently, CBM programs triangulated data across communities (20%) or used other types of validation (10%). Aside from validating data, most programs also measured the effort of the data collectors in their programs (66%).

Decision-making

A key purpose of monitoring is to influence decision-making. Involving people who face the daily consequences of environmental changes can enhance management responses across spatial scales and help in tackling environmental challenges at operational levels in resource management (Berkes and Armitage 2010). We assessed CBM programs in terms of their ability to influence decision-making at different levels, and found that they mainly informed decisions at a village (73%), regional (66%), or national level (40%). A small proportion of the programs also informed decisions at an international level (13%).

Scientific Support

We assessed the overall technical expertise that underpins the monitoring programs. Of the 20 CBM programs that responded, most had limited technical expertise available (65%). The programs generally had the necessary technical expertise to sustain operations, but some only had the expertise to sustain the program in its present state, not in the case of a major breakdown.

Sources of Funding

The majority of programs reported that they were being supported by government agencies (63%). A smaller number of programs were supported by civil society organizations (27%) and academic institutions (23%).

Sustained Funding

Access to stable, long-term financial support underpins the existence of CBM programs and ensures their long-term operation and sustainability. Most programs had no continuous support or expectation of follow-up funding (75%). Often, funding was tied to an individual project. Such structures can leave a large number of CBM programs vulnerable to changes in funders' priorities. A meager 15% of the CBM programs had a sustainable funding stream.

Long-Term Data Preservation

Contributions from CBM programs to broader repositories of Arctic observing data require consideration of data preservation practices. International standards suggest that an archive should keep more than one copy, use different media technologies, and preserve the datasets at different locations (Eynden et al. 2011; Parsons et al. 2011). Raw data and metadata should be retained to allow subsequent reprocessing (see further discussion in sections 3.5 and 3.6). Most CBM programs have no archiving system, or their data are only archived in a local archive. Frequently, long-term preservation of the data depends on a single small group of people (56%; n=18). One resource developed to support data on local observations and knowledge from Arctic communities is ELOKA (Exchange for Local Observations and Knowledge of the Arctic; https://eloka-arctic.org). While ELOKA was developed in the United States, it is available to the international community as a data repository.

Public Databases Derived from CBM Programs

Most of the CBM programs are presented on a website hosted by the program or the responsible institution (90%; Annex A). Typically, the websites present the aims of the program, the communities involved, and the attributes and locations that are being monitored. Occasionally, the CBM programs have made their data publicly available via these websites (47%; see also sections 3.2 and 3.5).

Overall Challenges

Programs reported that their principal challenge was limited funding (67%). Other challenges were participant turnover at the community level (33%), fatigue among community members (20%), various political challenges (17%), staffing turnover within government agencies (10%), and personal hardship (10%). None of the programs reported violations of intellectual property rights as a major challenge, presumably because they had effective ways of addressing the rights of Indigenous and local communities (see section 3.7). A more detailed assessment of the challenges and how to address them are provided in Chapter 4.

Sustainability

Most of the programs (67%) mentioned limited funding as the principal challenge to sustainable operations. As a proxy for the prospect of sustaining CBM programs, we assessed the financial costs related to data collection. The programs were surveyed on how much the equipment required by one data collector within the program would cost. In most CBM programs, the equipment needs were minimal (less than $100; 73%; n=30). In a few programs, however, the equipment cost was between $100 and $1,000 per data collector (20%), but could also be in excess of $1,000 per data collector (7%).

2.5   Linkages to International Agreements and UN Sustainable Development Goals

International Environmental Agreements

In response to the global environmental crisis, several hundred international environmental agreements have been adopted by countries around the world

(Mitchell 2003). These induce countries to change policies, and some have delivered major improvements by reducing environmental problems such as acid rain in Europe, the frequency of oil spills, the release of ozone-depleting gases, and reducing international trade in threatened wildlife (Kanie 2007). The Arctic governments (with a few exceptions) have adopted 10 international environmental agreements that are particularly relevant to the Arctic. The European Union has encouraged the implementation of these agreements in its Arctic policy (Table 2.3).

A major challenge in delivering international environmental agreements is linking the agreements to real-world decision-making. Jones et al. (2011) proposed that the most important objectives for monitoring progress in international environmental agreements were: (1) to evaluate environmental actions; (2) to inform policy choices; and (3) to raise awareness of sustainable development among the public and policymakers.

We assessed the international agreements in the Arctic and found that Arctic CBM programs could contribute to achieving the objectives of all 10 agreements of particular relevance to the Arctic (Table 2.3). Moreover, two stand out as being particularly suitable for contributions from the CBM programs: the United Nations Framework Convention on Climate Change and the Convention on Biological Diversity. A total of 100% of the CBM programs contribute to achieving the objectives of the UN Framework Convention on Climate Change, and 80% achieve the goals of the Convention on Biological Diversity. Our findings suggest that Arctic CBM programs have the potential to become very important vehicles for the implementation of environmental agreements in the Arctic.

United Nations Sustainable Development Goals

In 2015, the United Nations adopted the 17 Sustainable Development Goals (SDGs). There are 169 targets for the 17 goals. In its policy for the Arctic, the EU has attached particular importance to actions that are in line with the

---

Abbreviations: AEWA (Agreement on the Conservation of African-Eurasian Migratory Waterbirds); Bern (Convention on the Conservation of European Wildlife and Natural Habitats); CBD (Convention on Biological Diversity); CITES (Convention on International Trade in Endangered Species of Wild Fauna and Flora); CMS (Conservation of Migratory Species); ICRW (International Convention for the Regulation of Whaling); OSPAR (Convention for the Protection of the Marine Environment of the North-East Atlantic); POP (Convention on Persistent Organic Pollutants); Ramsar (Convention on Wetlands of International Importance); UNFCCC (UN Framework Convention on Climate Change).

Table 2.3. Multilateral environmental agreements in the Arctic that Arctic community-based monitoring programs could (blue), or probably could (orange), contribute to achieving (n=30 programs; Annex A).

Program	AEWA	Bern	CBD	CITES	CMS	ICRW	OSPAR	POP	Ramsar	UNFCCC
A-OK	■	■	■		■			■	■	■
Arctic and Earth SIGNs			■					■		■
Arctic Borderlands	■		■		■			■	■	■
Bird Phenology	■	■	■		■			■	■	■
BuSK			■							■
CSIPN	■		■					■		■
Evenk & Izhma Peoples			■					■		■
Farmers and Herders			■							■
Fávllis	■	■	■		■			■	■	■
FMI Snow Depth										■
Fuglavernd	■	■	■		■		■	■	■	
George River	■		■	■	■			■		■
Hares			■							
ION-Yukon River			■							
LEO	■		■					■		■
Marion Watershed			■							
Nordland Ærfugl	■	■	■		■		■		■	
Oral History	■		■							
Pilot Whale			■		■	■				
Piniarneq	■		■		■	■				
PISUNA	■		■		■	■		■		
Renbruksplan		■	■					■		
River Owners Iceland		■	■						■	
SIWO										■
Seal Monitoring			■			■		■		
Walrus Haulout			■		■			■		
Wildlife Triangles		■	■							■
Wild North		■	■		■		■			
WinterBerry										■
Älgdata		■	■							

SDGs in relation to research, science, and innovation on climate change, sustainable development, and international cooperation. We compared the questionnaire responses with the SDG targets and assessed which of the 17 SDGs the Arctic CBM programs contribute to achieving. We found that Arctic CBM programs contribute to achieving 16 of the 17 SDGs in the Arctic including "Life below water" and "Life on land" (Table 2.4).

It is likely that through improved data sharing and the wider use of digital platforms and global data repositories, CBM programs could also contribute to SDG 17, on technology partnerships, particularly with respect to elements of "enhance[d] knowledge sharing on mutually agreed terms" (Johnson et al. in review).

2.6 Summary: Capabilities

We examined the coverage, framework, and format of the knowledge products and their linkages to key international policies in the Arctic. Our assessment is based on an analysis of responses from 30 CBM programs that were chosen to reflect broad geographical coverage across the Arctic and a wide breadth of monitoring attributes.

CBM programs have existed in the Arctic for centuries. Our survey included several programs that started early in the 20th century, although most programs date to the early 2000s. Individuals of all ages, including children and elders, are engaged in Arctic CBM programs. Most CBM programs had less than 100 members, although there were a few that engaged with over 1,000 participants. While in general there is an overrepresentation of men in CBM programs, there are also programs that mostly involve women. Since many CBM tasks involve observing environmental change, programs may disproportionately engage with men who are fishers, hunters, and herders who frequently move around and spend a lot of time observing the environment. However, more research is needed on what types of community tasks are involved in CBM programs to understand the underlying gender differences.

Community members participate in CBM programs for many reasons. Important motivation for community involvement in CBM includes being visible and recognized contributors in the local democracy, being a part of local decisions, and being able to protect one's rights to local resources. Another important motivation is to be involved in the process of decision-making to sustain the health and abundance of wildlife resources upon which the communities depend. Many of the successful CBM programs reported that they were contributing to capacity enhancement of the local community in multiple ways: socially, politically, cognitively, and economically.

Table 2.4. The proportion of Arctic community-based monitoring programs that contribute to achieving the United Nations Sustainable Development Goals (n=30 programs; Annex A).

Most of the CBM programs in our survey focused on the observing domains of "land/cryosphere" and "ocean/sea ice." CBM programs use methods from both the natural and social sciences. This is to be expected, given that Arctic CBM programs monitor a wide range of phenomena, including biological attributes, abiotic phenomena, and sociocultural attributes. By nature, CBM programs tend to focus on issues that are of greatest concern to local stakeholders, and they prioritize observing those attributes.

Focused observations on community priorities also means that CBM programs have considerable potential to influence on-the-ground management activities. CBM can complement scientist-executed monitoring and contribute to the broader Arctic observing landscape by enabling an increase in sample size, sampling density, areas, and time. Most Arctic CBM programs make observations between 61 and 70° N. Observations are undertaken in all major

Arctic biomes, with the exception of ice desert; low observation coverage occurs in areas rarely visited by community members. Many CBM programs allow data collection to occur year-round.

The majority of Arctic CBM programs in our sample involve Indigenous knowledge. CBM programs inform decisions at village, regional, and sometimes national levels. We found that CBM programs contribute to better-informed decisions or better-documented processes in six key economic sectors in the Arctic: hunting/herding, forestry, shipping, fisheries, tourism, and mineral and hydrocarbon extraction. It is often governmental agencies that fund the work of the Arctic CBM programs.

We explored the topics of data quality, data access, and use, and found that data from individual CBM programs are typically stored in only one repository. This makes the data highly susceptible to loss. Some CBM programs have made their data publicly available, but only a few programs are connected with data discovery portals or global repositories.

Finally, we assessed if CBM capabilities could be used to support the goals of 10 international environmental agreements relevant in the Arctic and the 17 UN Sustainable Development Goals. We found that Arctic CBM programs could contribute toward all 10 international environmental agreements, with the majority providing relevant information for the United Nations Framework Convention on Climate Change and the Convention on Biological Diversity. Moreover, we found that CBM contributed to 16 of the UN Sustainable Development Goals.

Despite their potential value for meeting international environmental objectives, most Arctic CBM programs considered sustained funding to be a significant challenge, even though many programs operate at very low costs. Other important challenges in sustaining CBM programs included participant turnover at the community level and fatigue among community members. Learning from best practices to overcome these and other barriers is key to future CBM sustainability. We address these issues with examples of successful cases in Chapters 3 and 4.

3 Good Practice

Few people, extreme environments, huge distances, limited bandwidth, communities with a very close and often historic connection to the environment—these are some of the features that characterize the Arctic.

How do you undertake CBM programs under these conditions? In this chapter, we discuss good practice in Arctic CBM programs in terms of establishing, implementing, and sustaining programs; obtaining impacts; being relevant and meaningful; and connecting with other approaches. We also discuss how to ensure the quality of monitoring data and knowledge products, and address the rights of participating Indigenous and local communities. Further suggestions for good practice in Arctic CBM programs are available in Parlee and Lutsel K'e Dene First Nation (1998), Ecological Monitoring and Assessment Network (2003), Gofman (2010), Sigman (2015), and Johnson et al. (2015, 2016).

This chapter is based on a review of the scientific literature and workshops with CBM program practitioners and community members engaged in CBM programs. The workshops were held in Nuuk (December 2016), Fairbanks (May 2017), the Russian districts of Komi Izhma, Zhigansk, and Olenek (September 2017, September 2018, April 2019), Québec (December 2017), and Longyearbyen (March 2019); see Table 1.1.

By "good practice" activities, we refer to activities that CBM program facilitators or community members engaged in CBM programs have highlighted to us as being important and effective when undertaking CBM programs or activities that have been identified as useful in the scientific peer-reviewed literature.

The most recent review of good practice in Arctic CBM programs was based on workshops held in 2013 and 2014 (Johnson et al. 2016). There have been several important developments since. At the policy level, the use of Indigenous and local knowledge for informing decision-making has received increased attention. In 2016, US President Barack Obama convened the first Arctic Science Ministerial, meeting with Arctic Indigenous leaders to get their input into priorities for Arctic science. The joint statement released after the ministerial mentions community-based observing and "traditional and local knowledge" as components of an integrated Arctic observing system (White House 2016a). The Canadian government announced its intention to co-develop a new Arctic policy framework with northerners and Indigenous peoples (White House 2016b). Increased interest in Indigenous perspectives is also now seen in some national Indigenous districts in Russia (e.g., in the Evenk national Indigenous district declared in Zhigansk, Yakutia). Another example is the Greenland government's coalition agreement in 2016 that specified at the ministerial level that the government's resource management should be based on both scientific advice and the observations and knowledge of Indigenous and local community members. Moreover, in 2018 the hunters and fishermen engaged in the CBM program Piniakkanik Sumiiffinni Nalunaarsuineq (PISUNA) in Greenland were awarded the prestigious Nordic Council Environment Prize. Also in 2018, the first binding agreement on integration of Indigenous/local knowledge with science knowledge for Arctic resource management was signed ("The Agreement to Prevent Unregulated High Seas Fisheries in the Central Arctic Ocean"). At the technical level, there has been rapid development and uptake of user-friendly approaches to storing and sharing data and information in CBM programs (Johnson et al. in review).

In this chapter, we highlight examples of good practice in Arctic CBM programs while emphasizing emerging approaches that might be of broad interest.

3.1 Good Practice: Establishing CBM Programs

New CBM programs are established in the Arctic every year (Kouril, Furgal, and Whillans 2015). Careful design can go a long way toward making the programs successful.

CBM programs must address the communities' priorities, which means that communities need to define the questions. Communities should be informed about plans, threats, risks, and challenges concerning local communities and resources known to the regional and national level, and be invited to collaborate to engage in and bring knowledge concerning these issues. Moreover,

Photo by local observer to document the abundance of the coastal stock of Atlantic cod (*Gadus morhua*) in Disko Bay, Greenland. The photo was taken off Akunnaaq in September 2010. Credit: Gerth Nielsen, PISUNA

the communities themselves should define the approaches for measuring and answering the questions (Fairbanks Workshop; Cochran et al. 2013).

Community meetings to introduce and discuss the idea of a local CBM program should be in the form of a dialogue. The focus should be on fruitful cooperation, and the role of the facilitators should primarily be to listen to the community members. The facilitator should be mindful of new ways of looking at things and avoid preconceived conclusions. Participants must come to their own conclusions, together, following the discussions.

How CBM programs are established in practice depends on the context—particularly the stakeholders and the land/resource tenure system—as well as the proposed objectives of the CBM program. A number of CBM programs address fishing, hunting, and herding, which are key economic sectors and important to the identity and culture of many Indigenous and local communities in the Arctic. An example from Arctic Russia is described in Box 3.1.

During the design of CBM programs, one needs to consider what questions the monitoring aims to answer, and what community members and others are likely to do with the results. Moreover, management relations and support structures (roles and responsibilities, communication lines, funding)

BOX 3.1. EXAMPLE

Introducing a CBM Program in Arctic Russia

CBM PROGRAMS OFTEN BEGIN WITH A MEETING with the local community to introduce and discuss the idea. Prior to this meeting, the community leader must have identified the local fishers, hunters, and herders—men and women—who are particularly knowledgeable about the natural resources.

At this initial meeting, the facilitator typically asks the participants about the challenges they face. Some of these challenges are often related to their livelihoods of fishing, hunting, and herding. Community-based monitoring is then introduced as a tool that can assist with documentation and communication of changes or challenges to decision-makers. This approach can be useful in supporting communities to address challenges in a proactive way.

As an example, within a program in Arctic Russia led by the Centre for Support to Indigenous Peoples of the North, the Evenk and Izhma participants found that they wanted to monitor (workshops in the Russian communities Zhigansk and Komi; Enghoff et al. 2019):

- Fish species, fishing activities, and fishing methods.
- The animals that the community members hunt.
- Attacks by predators.
- The use of resources in the area by people from inside and outside the community.
- The changes in climate and the environment around the community (snow, ice, pollution).

When establishing this CBM program, the facilitator agreed with the community members that they should regularly collect and share their observations on these topics. This could improve the livelihoods of the community and strengthen the community's rights to the use of the resources. More specifically, it would contribute to:

- Improved and more sustained access to fish.
- Better hunting regulations for animals that are hunted.
- Better management of predators.
- Better acknowledgment of the rights of their own community to use their land.
- Improvements in addressing pollution in rivers and lakes.

In CBM programs of this kind, monitoring is usually done as part of the routine fishing, hunting, and herding activities. It is important to provide community members with guidance in terms of field techniques and how to make use of the data. Members of the CBM group will observe the various aspects of the environment in their area and meet regularly—for example, every three months—to discuss trends in natural resources and possible reasons for changes. Management actions can be proposed and then communicated to the local authorities or to the local interest organizations for the Indigenous peoples.

also need to be considered (section 4.6). If the purpose of the CBM program is to influence decision-making, an understanding of who makes decisions needs to be obtained along with agreement on how the monitoring results will reach this body (section 3.4). Ideally, there should be mechanisms in place for program participants to interpret the data, as well as protocols for how the decision-making body will use this data to inform decisions. This may help prevent alarmist reactions such as prohibitions on resource use based on limited understanding of data, as well as minimizing the potential for decision-making bodies to ignore data contributed from CBM programs.

Sustainability of the effort should be incorporated into the program design from the start if the goals require long-term work. One important approach to enhance sustainability of the observations is to respect the existing political and organizational structures in the area, and embed the monitoring within these existing institutions (Québec Workshop). Another important way is to keep the methods as simple and locally appropriate as possible (sections 3.3 and 4.3), for example, by aligning monitoring with routine activities such as hunting and harvesting.

Men and women often have different community perspectives. For instance, men may focus interest on the weather and access to resources, and this frequently gains more attention. However, women may focus on the health of the plant, fish, or animal by observing the quality of meat and the changes in the berries (Fidel et al. 2017). Depending on their role, a person will have different interests and needs, and so will be focused on different aspects of the environment (e.g., agriculturalists may focus on rain while herders may focus on freeze and thaw cycles). In Indigenous communities in the Arctic, elders and youth often have different perspectives and strengths—for example, related to Indigenous knowledge and technology use—but may be equally interested in contributing to a monitoring program. To ensure that CBM programs address a community's priorities, a diverse array of people representing the community should be directly involved in program planning and design (Fairbanks Workshop). This diversity should not be prescribed by partners from outside the community but should be discussed openly and agreed upon by all involved.

Conclusions: Establishing CBM Programs

Community members should play a central role in planning CBM programs. The monitoring should reflect the local priorities, with attention to diversity within communities (including men's and women's priorities), and be kept as simple and locally appropriate as possible. Management relations and support structures should be carefully considered during the design of CBM programs. The programs should be embedded within existing organizational structures.

It is important (but not easy) to set up procedures to ensure that the community members' observations and management proposals reach and can be used by the management authorities in the area (see below). Attention should be paid to providing guidance in field techniques to community members; community members and decision makers would benefit from guidance in how to interpret and make use of the data.

3.2 Good Practice: Implementing CBM Programs

Moving from establishing to implementing CBM programs will require organizational and support structures to sustain the effort.

Effective communication is critical to implementation of all types of CBM programs involving partners from outside the community, such as scientists or management agencies or other decision makers (Fairbanks Workshop). In programs aimed at informing decision makers, maintaining interest among community members hinges on effective two-way communication between the participants and the management agency (discussed further in section 3.3). At the community level, at least one person needs to have responsibility for ensuring that meetings are held and that community members' observations are discussed and communicated to the management authorities, and that feedback from the management authority is shared with all the participants. There is a risk that if the management authority does not take action or respond in a timely manner to the reports they receive, the community members may lose interest (Québec Workshop). The CBM program must encourage the management authority to take the government policies of listening to local stakeholders seriously and ensure proper follow-up of information provided by community members. Establishing parallel reporting lines can sometimes reduce delays in communication (Nuuk Workshop).

Common ownership and shared goals and values at local level are likely to make it possible to collect more data, get in-depth information and broad knowledge, and to develop projects without too many time-consuming conflicts. CBM programs can thus contribute to better results, a tailor-made monitoring approach, and increased innovation capacity. In order for co-creation to be possible over time, it is important to base the collaboration on mutual trust and respect, and the understanding of the time and work needed for data collection to be sustainable over time, without wearing out the actors involved at local level (see also discussion in section 4.6).

Observing events in nature influences reindeer herders' survival strategy and resource use. Through CBM programs, their observations can also contribute to the wider understanding of the changes in the Arctic. Nenets Autonomous Okrug, Russia. Credit: Finn Danielsen

Digital communication tools, including social media applications, can be useful for exchanging observations, ideas, and experiences between community members, further enhancing their commitment and engagement (examples below) (Fairbanks and Québec Workshops; Johnson et al. in review). In the same vein, it can be very useful for participants to meet community members engaged in other CBM programs, to share information and gain inspiration. This may be done by organizing direct exchanges between CBM programs or by hosting workshops or other types of convenings.

The existing CBM programs provide many examples of how data and information are stored, and results are fed to users and back to data gatherers. Sometimes CBM programs store their data and information in ring binders or Excel spreadsheet files on a computer. The data can be located in the community with one of the community members or in the office of the institution facilitating the CBM program. One advantage of using ring binders is that they are easy to maintain over time. The archiving and retrieval process is not dependent on external assistance, and recurrent costs are minimal (Brammer et al. 2016). However, a distributed archiving process can help with data backup and reduce the risk of losing data stored in a single location.

Feeding results back to the data gatherers often requires special efforts. In Finland, Snowchange Cooperative uses videos for reporting the results of oral history documentation of Indigenous and local knowledge about river basins, reaching community members, youth, and the broader public (Mustonen 2014; Mustonen et al. 2018). The University of the Faroe Islands stores data from hunters' self-monitoring of mountain hare *Lepus timidus* hunting in a web-based database and uses Facebook to discuss managing the outfield landscapes with the hare hunters (Danielsen et al. 2017; Magnussen 2016; haran.fo).

One disadvantage of storing CBM data and information in ring binders and digital files is that the data can be difficult to access, not only for other community members but also for decision makers and those who can influence management decisions. This problem gets worse when data and information are accumulated from wider areas and over a longer time span. Some CBM programs use web-based digital (knowledge management) platforms for storing data and for feeding the results back to data gatherers. Several CBM programs have developed their own platforms. The Gordon Foundation has set up a platform for sharing data on water quality (MackenzieDataStream.ca). This platform is used in 23 communities along the MacKenzie River. With support from Google, the Arctic Eider Society has developed a platform that connects with social media networks (siku.org/about). This is used for multiple Inuit-identified priority areas within research, education, environmental stewardship, and health in five communities in Hudson Bay (Québec Workshop).

The Geomatics and Cartographic Research Centre at Carleton University, with partners such as the Kitikmeot Heritage Society and Exchange of Local Observations and Knowledge for the Arctic (ELOKA), are cooperating with several CBM programs to lead the development of an adaptable open source software called Nunaliit (nunaliit.org) to develop tailor-made platforms for visualizing and storing different types of data and information generated by CBM and community-led research projects. Platforms include digital atlases and maps and online databases that store, archive, and share observations documented by Indigenous observers, with attention to ensuring that communities control data access.

ELOKA uses a series of platforms to support CBM, including Nunaliit and custom products such as the SIZONet platform (eloka-arctic.org/sizonet, eloka-arctic.org/pisuna-net). The Yup'ik Environmental Knowledge Project and Atlas (eloka-arctic.org/communities/yupik/atlas/index.html) shares and documents Yup'ik place names and environmental knowledge in the Yukon-Kuskokwim Delta in Alaska under the guidance of Calista Education and

In the Faroe Islands, results from hunters' self-monitoring are fed back to the data gatherers with the use of Facebook. Credit: Eyðfinn Magnussen

Culture. The online atlas contains over 4,000 Indigenous place names and a rich collection of stories, videos, and other related information. The communities have trained local students to add their own data to the atlas (Québec Workshop).

The Government of Nunavut's Department of Environment documents Inuit knowledge through participatory mapping of fish, marine mammals, birds, aquatic plants, and invertebrates. They have completed the mapping of 22 coastal communities. This program feeds data into an online map-based platform (Nunaliit) and, at the same time, prepares paper reports with monitoring results for the communities. The goal is to be able to update the maps over time. The role of the platform is to help direct science priorities, identify fisheries development priorities, and advise management bodies (Québec Workshop).

The Local Environmental Observer network of the Alaska Native Tribal Health Consortium has established a platform where individuals can post observations when they see something unusual, for instance related to seasonality, timing of snow melt, water quality, or sanitation (leonetwork.org). Community members who post an observation are often connected to a scientist who can answer questions about the issue. This program keeps observers engaged through a digital newsletter and also hosts conference calls. The network also serves as a link to more systematic community-based monitoring, such as harvest monitoring by local government agencies or sea-ice observations that feed into ELOKA (Fairbanks Workshop).

Conclusions: Implementing CBM Programs

Online platforms create possibilities for sharing community-produced observations across sites and scales of decision-making. Some disadvantages, however, include the expertise and resources required to establish and maintain a platform, and poor connectivity in rural communities limiting community access to data. Because funding is often time-limited, platform maintenance is a particular challenge. Programs should consider an appropriate long-term repository of data, including the best platform or data repository, during the design phase. The discussions should consider implications for community access and control of data and information beyond the period of available program funding.

3.3 Good Practice: Sustaining CBM Programs

To sustain CBM activities and ensure that programs continue to be relevant over time, it is important to maintain the long-term interest and involvement of community members. When a management agency is involved or leading the process, it is also important to avoid frequent staff turnover.

What can CBM programs do to maintain long-term community interest and involvement? Several CBM programs suggest the strongest incentive is seeing that the information from the observing efforts is actually being used to inform management decision-making (further discussed in section 3.4). CBM programs must ensure that information is made available to decision makers at the relevant and appropriate scale, including community organizations (such as Hunters and Trappers Organizations in Canada), co-management boards, and regional and state agencies when appropriate (Nuuk Workshop).

Another opportunity is to use tools and approaches for data collection that can be easily incorporated into a community's day-to-day activities (section 3.1). CBM programs can also consider carefully how program participants benefit from the program and set up a reward system of tangible benefits for community members who participate. In some parts of the Arctic, including Canada and Alaska, there is growing agreement that community observers should be offered monetary compensation, often in the form of stipends or

A youth in Sitka, Alaska monitors one of their adopted *Empetrum nigrum* (crowberry or black crowberry) plants that they named "French Fry" during the University of Alaska Fairbanks Winterberry Project. Volunteers name their plants to establish a relationship and to keep their data better organized, and make weekly counts of the berry abundance and condition (ripe, rotten, dried out, damaged). Photo by Jasmine Shaw, UAF.

gas payments for active harvesters, as well as creating paid coordinator positions (Québec Workshop). In other regions, tangible benefits may include a small symbolic honorarium for lost work time. A negative aspect of monetary compensation is that it may make it difficult to sustain the program over time outside of an externally funded project context.

In Nunavut, the Community-Based Monitoring Network of the Nunavut Wildlife Management Board collects data on observations and hunting activities based on the 1992 Land Claims Agreement (Etiendem et al. 2020; www.nwmb.com). The network has a goal of providing information to the board that will help develop management plans; identify important harvesting areas; document species distribution, movement, and health of wildlife; and identify issues that may require further research. A clerk in each community uploads the observation data such as start and end point of each hunting trip, means of transport, and what was harvested. Different cash or prize drawings are made every other week, monthly, and twice a year as incentives for observers in participating communities (Québec Workshop). After three full years of data collection in three communities, the program moves on to three new communities, based on a call for expressions of interest from other Nunavut communities interested in participating in the network.

Frequent staff turnover at the management authority level can constrain sustainability, and this can be difficult or impossible to avoid. Encouraging the involvement of multiple management authority staff members reduces the risk that loss of any one person will have a detrimental effect on the CBM program's overall activities (Nuuk Workshop).

Conclusions: Sustaining CBM Programs

CBM programs can encourage continued interest among community members by ensuring that the participants' observations are used for decision-making and that they are informed of how the information is being used. They can also use tools and approaches for data collection that can easily be incorporated into participants' day-to-day activities, and set up a relevant reward system. The effects of frequent management authority staff turnover can be minimized by involving multiple staff members in the CBM program.

3.4 Good Practice: Obtaining Impacts Through CBM

Arctic CBM programs have impacts on a wide range of decisions, from quotas and fishing permits to wetland restoration, Indigenous food systems, and safety measures (see below).

Many CBM programs have developed and begun using specific protocols and procedures to enable management authority agencies to incorporate local and CBM-derived information on natural resources and resource use in decision-making. Nevertheless, CBM practitioners report that management agencies are slow to take action in response to observations emanating from CBM programs. This suggests that further advocacy is needed to get management agencies to follow government policies on local stakeholder participation in land and resource management, and to incorporate information from CBM programs into their decision-making (discussed in section 4.1). This also suggests that further efforts are needed to raise awareness within management agencies as to the value of Indigenous and local knowledge and observations, and to wholeheartedly incorporate CBM programs into national policies, the administrations' funding streams, and the job descriptions of the Arctic governments (Nuuk Workshop).

In Iceland, the Federation of Icelandic River Owners has tracked the number of salmon and trout fished in 180 rivers across the country since 1932. Every year more than 1,000 anglers report their catch. The results are used to divide the economic benefits from the river owners' selling of fishing permits and to ensure sustainable harvesting (angling.is). The monitoring program operates in accordance with the Fisheries Act of the Government of Iceland (No. 61 of 2006, with adjustments). With national policy support, the program has been institutionalized, and today constitutes a central component of Iceland's management and monitoring of its rivers.

In Greenland, the Piniakkanik Sumiiffinni Nalunaarsuineq (PISUNA) program of the Ministry of Fisheries and Hunting introduced a system for communities to advance natural resource management recommendations to municipal and national authorities based on the fishers' and hunters' own review and assessment of observed trends in the status of natural resources. PISUNA tracks management interventions resulting from the monitoring. Ninety interventions have been proposed involving 30 species of fish, mammals, and birds. Some proposals have been adopted (e.g., changes in hunting seasons of eider ducks and Canada geese [*Branta canadensis*]), but many are awaiting action by the local or central authorities. The proposals are presented to the local government authority by experienced fishers and hunters organized in local natural resource councils. Moreover, the proposals are published at www.pisuna.org and at PISUNA-net, a searchable database (eloka-arctic.org/pisuna-net/). These experiences suggest that management interventions are useful to track because they can indicate the possible management impact of the program and provide direction about the aims of the CBM program (Nuuk and Fairbanks Workshops).

In Canada, the Arctic Borderlands Ecological Knowledge Society monitors and assesses ecological changes within the range of the Porcupine caribou herd and adjacent coastal and marine ecosystems. Community monitors document information during interviews with local experts. The Porcupine Caribou Management Board and the government use this information for co-management and decision-making, particularly with regard to caribou harvest quotas for Gwich'in and Inuvialuit harvesters. Researchers also rely on this information (Québec Workshop).

In Finland, Snowchange Cooperative has established a CBM program with fishers in Jukajoki Basin, a peat swamp that has been heavily damaged by industrial activities and is the site of one of the largest aquatic habitat restoration activities in the Arctic (Mustonen 2014). The CBM program helps fishers document the fish resources and communicate key aspects of local knowledge

Changes in the environment have direct influence on the lives and livelihoods of Arctic communities. If policies support CBM, CBM programs can impact a wide range of decisions, from quotas and fishing permits to wetland restoration, Indigenous food systems, and safety measures. Credit: Martin Enghoff

about the peat swamps—for instance, through videos (see section 3.2)—to those in charge of restoring the wetlands (snowchange.org). The program thus helps guide the habitat restoration efforts, contributes to co-management, and complements other initiatives to include local perspectives in the stewardship of the landscape (Nuuk Workshop).

In Norway, the International Centre for Reindeer Husbandry has tested the use of Indigenous and local knowledge for improving Indigenous Saami food systems. Their perspective is to develop Indigenous and local knowledge standards for improved production and processing of reindeer-based food. If the standards can be incorporated into national legislations, Saami communities could enter mainstream markets with their products. They have so far documented reindeer herders' Indigenous knowledge standards on Saami slaughtering processes and meat-smoking practices (reindeerherding.org) (Nuuk Workshop).

In Alaska, the Sea Ice for Walrus Outlook (SIWO) initiative provides weekly reports from April through June, depending upon presence of sea ice, on weather and sea-ice conditions relevant to walrus in the northern Bering Sea and southern Chukchi Sea regions of Alaska. The National Weather Service contributes satellite images and ice and weather forecasts for each community. Local observers send in their observations and pictures of the current conditions each week, as well as relating similarities and differences compared to years past, helping improve operational weather and ice forecasts (Deemer et al. 2017). While this collaboration improves National Weather Service products, the data are not specifically used to inform governmental decision-making. The SIWO information is posted to the project's website and Facebook page. Hunters in Indigenous communities can use the information to plan and coordinate scouting and hunting trips (Fairbanks Workshop).

Conclusions: Obtaining Impacts through CBM

Arctic CBM programs can impact many kinds of decisions by providing information to management authorities and community members. Some CBM practitioners find it useful to track the management interventions that result from CBM programs. Greater impacts can be obtained by further developing protocols and procedures to enable management agencies to incorporate CBM-derived information into decision-making, and by bringing communities together, sharing information, and promoting advocacy on the importance of using information from CBM programs. Greater impacts can also be achieved by further developing national policies in support of CBM programs and requirements to incorporate information from CBM into decision-making processes.

3.5 Good Practice: Connecting and Cross-Weaving with Other Approaches

As well as providing data with which to inform management decisions at local or provincial and national levels, CBM has the potential to shed light on changes in the environment on a national and even international scale (IASC 2013). In this section, we will highlight good practice among Arctic CBM programs in (1) connecting with scientist-executed monitoring programs and (2) contributing to global data repositories to ensure maximum scientific usage and track larger-scale trends.

In terms of CBM programs connecting with scientist-executed monitoring programs, there is limited knowledge of good practice. Some Arctic CBM programs are intertwined with scientist-executed monitoring programs through the initial interpretation level (communities contribute to designing the methods and to collecting data and initial interpretation of the results; the preliminary results are then handed over to scientists), and scientists present the (combined) results for decision makers. Other CBM programs run on their own: representatives of community members or facilitators of the CBM programs present the results to decision makers, who may receive information on the same topic from parallel monitoring programs undertaken by scientists (in those few cases when such information is available).

One CBM program that is successfully intertwined with larger-scale scientist-executed monitoring is the Indigenous Observation Network (ION), which is led by the Yukon River Inter-Tribal Watershed Council and includes partner Indigenous communities and the US Geological Survey. Data collected by ION is informing basin-wide planning of the Yukon River in Alaska and Canada. Communities have used the data for community planning on safe drinking water, wastewater, and solid waste management, and for advocacy to protect clean water and salmon stocks. Regionally, the data have been used to inform the Yukon River Water Quality Plan that was adopted by Alaska Native Tribe and Canadian First Nation representatives in 2013. The plan aims to keep the Yukon River and its tributaries "substantially unaltered from natural conditions," and the data generated by the ION is used to guide further monitoring efforts and assess river conditions with this overarching goal in mind. Other examples of CBM programs that are intertwined with larger-scale scientist-executed monitoring are the Alaska Arctic Observatory and Knowledge Hub (section 3.7) and the Federation of Icelandic River Owners (section 3.4).

Examples of CBM programs that run on their own and forward results to management authorities for decision-making after comparison with advice from scientists include fishing and hunting harvest statistics programs in many countries, and the Piniakkanik Sumiiffinni Nalunaarsuineq program in Greenland (sections 3.4 and 3.6).

In terms of CBM programs connecting with international repositories, there are many open-access global repositories that may be of relevance to Arctic CBM programs. Thirteen examples of repositories from the federated metadata search portal, the Arctic Data Explorer (nsidc.org/data), are listed in Table 3.1. According to this website, a few CBM programs have already shared datasets with these repositories or are referenced by their catalogues. A total of 113 out of the 42,946 datasets in these 13 repositories are derived from CBM programs (0.3%).

Moreover, to help make the datasets derived from monitoring programs discoverable, data discovery catalogues may be useful. For instance, 46 monitoring programs with CBM or citizen science components in the Arctic,

Many observations of the environment are made by local observers in Arctic CBM programs. Only a small proportion of the datasets have so far been included in international data repositories. Credit: Martin Enghoff

including the Western Arctic Beluga Health Monitoring and the Inuvialuit Settlement Region Community-Based Monitoring Program, have published metadata on their datasets in the data discovery catalogue (eudatmd1.dkrz.de/dataset?tags=Community-based+monitoring).

CBM programs may potentially fill gaps in the existing Arctic data delivery chains. Our overview of attributes monitored by the existing CBM programs (Chapter 2) suggests that they are particularly effective in tracking the delivery of goods and services from natural ecosystems. Ecosystem benefits are a prime focus of several international environmental agreements (e.g., Convention on Biological Diversity 2019), yet are extremely hard to monitor using a top-down approach. Moreover, CBM programs can complement Arctic scientist-executed monitoring programs by enabling an increase in sample size, area, and time. This, however, requires increased collaboration between scientists and CBM programs, learning from the existing "good practice" examples of connected approaches, and targeted investments (Eicken et al. in review).

Table 3.1. Examples of open access global repositories relevant to Arctic CBM programs and the number of datasets derived from Indigenous and local knowledge. Indigenous and local knowledge data may only include metadata, pointing to repositories such as ELOKA. Source: arctic-data-explorer.labs.nsicd.org (Dec. 2019).

Repository	Datasets total	Datasets derived from Indigenous and local knowledge
National Science Foundation Arctic Data Center	10,753	48
Polar Data Catalogue	2,557	46
National Aeronautics and Space Administration Earth Observing System Clearing House	14,316	12
US Geological Survey ScienceBase	492	7
National Oceanic and Atmospheric Administration, National Oceanographic Data Center	6,352	0
NOAA's National Centers for Environmental Information, World Data Service for Paleoclimatology	3,703	0
Biological and Chemical Oceanography Data Management Office	1,526	0
Global Terrestrial Network for Permafrost	1,404	0
Rolling Deck to Repository	914	0
International Council for the Exploration of the Sea	470	0
Norwegian Meteorological Institute	201	0
University Corporation for Atmospheric Research, National Center for Atmospheric Research, Research Data Archive	173	0
The Digital Archaeological Record (tDAR)	85	0

Participants at the Fairbanks Workshop stressed the immense value of data sharing. The more information is distributed, the more valuable it becomes because people can use it. While there is growing emphasis on data sharing in the global research community—reflected, for example, in the FAIR principles of being findable, accessible, interoperable, and reusable (Wilkinson et al. 2016)—participants also noted that not all data can or should, be shared (Pulsifer 2015). Intellectual property rights, data sovereignty, and customary law must be respected (Scassa et al. 2015; Young-Ing 2008). Data sovereignty is the idea that data are subject to the laws and governance structures within the area they are collected. The Global Indigenous Data Alliance released the CARE principles (collective benefit, authority to control, responsibility, and ethics) as a companion to the FAIR principles to emphasize the specific data-management requirements of Indigenous knowledge (www.gida-global.org/care).

CBM programs are based on Indigenous and local knowledge systems that cannot be directly compared with scientist-executed monitoring programs. When connecting different branches of science with CBM programs, one has to recognize the often asymmetric power issues arising (Tengö et al. 2017; Tengö et al. in review). When scientists "assimilate local ecological knowledge within Western worldviews" (Mistry and Berardi 2016), there is a risk that it may further marginalize Indigenous and local people (Latham and Williams 2013). For organizers of CBM programs to effectively share their data with global repositories, suitable terms of cooperation must therefore be established. Agreements on cooperation between CBM programs and the global repositories should address principles of intellectual property rights and Free Prior and Informed Consent (further discussed in sections 3.7 and 4.5; United Nations 2008), respect for knowledge holders, and reciprocity (Pulsifer et al. 2011).

Conclusions: Connecting and Cross-Weaving with Other Approaches

There is limited knowledge on good practice in connecting CBM programs with scientist-executed monitoring programs. Some programs are intertwined with scientist programs at the interpretation level; others run independently and in parallel. Further work is required to identify the gaps in existing Arctic data delivery chains that CBM programs might plug into. Examples of the successful incorporation of both CBM and scientists' program data into decision-making should be highlighted to encourage further cooperation. Only a tiny number of CBM datasets are currently included in Arctic data repositories. Representatives of Indigenous and local communities should decide whether data from their CBM programs should be connected with global repositories, and the process must take place in accordance with their free, prior, and informed consent. When appropriate, CBM programs could make their datasets publicly available and connect with global repositories founded for the purpose, such as ELOKA.

3.6 Good Practice: Ensuring the Quality of Knowledge Products

One barrier to maximizing the potential of Indigenous and local knowledge in CBM programs for decision-making is the perception among management agency staff and scientists that information from local people is subjective and anecdotal (Eicken, Ritchie, and Barlau 2011; Moller et al. 2004). A growing body of literature demonstrates, however, that where Indigenous and local knowledge is systematically gathered, data collected by community members are as reliable as those of professional scientists (Danielsen et al. 2014a, 2014b, 2014c; Herman-Mercer et al. 2018).

The community perspective is relevant too. If scientists do not possess the social and cultural skills to appreciate context and locality, then Indigenous and local communities will view their initiatives with suspicion. There is therefore a need to establish credibility in both directions (Huntington 2011; Huntington et al. 2013). Describing and discussing good practices in how scientists can improve their legitimacy and credibility within communities, however, is beyond the scope of this report.

In this section, we draw on information shared in the CBM workshops and previous research on CBM internationally to discuss good practices in terms of accuracy and precision of knowledge products from CBM programs. First, however, it is important to acknowledge that wildlife population estimates based on community observations have at times been more accurate than scientific estimates (see, for example, Albert 2000). Moreover, it has been shown that Indigenous community members and scientists sometimes observe different segments of the same population of wildlife. In Australia, Indigenous community members detected wildlife individuals that were shyer and more difficult to see than those detected by scientists, suggesting that one of the groups' sampling was unrepresentative (Ward-Fear et al. 2019). Although measurements by community members can compare well with similar measurements by scientists, CBM approaches to monitoring can, in some contexts, be more vulnerable than professional techniques to sources of bias, thus decreasing their accuracy (defined as the closeness of the resulting measurements to their true values).

Key potential problems include a lack of measuring experience on the part of observers (which may lead to over- or underestimates of abundance

Local observer Joe Leavitt explains features of sea ice to students and scientists in Utqiagvik, Alaska. Credit: Hajo Eicken

and size); potential conflicts of interest (with recorders perhaps inadvertently providing data that are biased toward managers' preconceptions); a tendency, in the absence of careful documentation, for methods to drift over time, or for results to reflect long-term ("fossilized") perceptions more than current trends; and the potential for the spatial or temporal coverage of monitoring to be unrepresentative of the entire system of interest (Danielsen, Burgess, and Balmford 2005).

Besides accuracy, the utility of monitoring can be limited by the precision of the results (that is, the closeness of repeated measurements of the same quantity to each other). Sources of low precision (leading to high variance around the estimated true value of the attribute of interest) may include small sample sizes; overly thin or patchy temporal or spatial deployment of sampling effort; the physical loss of data; and the inconsistent application of methods, either over time or across observers. These problems can affect all monitoring but are likely to be a particular problem where financial or professional human resources are limited.

The potential limitations of CBM can be overcome by careful planning, explicit consideration of likely biases, and thorough guidance, training, and quality control checks.

CBM programs use sample- and perception-based methods for data collection (further discussed in section 2.4). For sample-based programs, one way of reducing bias is to allocate individuals with a rank based on their knowledge, allowing data to be disaggregated according to this ranking. Different community members are likely to have more or less expertise in different topic areas. Community members are often most capable of ranking other community members' expertise. Biases can be reduced by discussing and interpreting disaggregated data at community meetings, where feedback can be obtained from the entire community. For perception-based CBM programs, other measures are relevant, as summarized in Table 3.2.

What can scientists do? Scientists could allocate some time and effort to visit CBM programs and learn about them. Moreover, scientists could clarify to themselves the potential contribution of CBM programs to their own research, and how they can guard against data quality issues in their research. A program to fund scientist visits to consolidated CBM programs might help. In addition, funding institutions and governments could request collaboration with local communities and existing CBM programs when developing new funding calls for research projects. This would contribute to UN Sustainable

Table 3.2. Measures that CBM programs can take to increase their ability to provide natural resource abundance data that trained scientists would consider reliable (Danielsen et al. 2014b; see also Huntington 1998).

1.	Establish independent focus groups in multiple communities that know resource abundance in the same geographical area (triangulation across communities).
2.	Convene regular (e.g., annual) village meetings to present and discuss data and interpretation and obtain feedback from the entire community (triangulation across community members).
3.	Facilitate the collection of auxiliary data through, for example, community members' direct counts of resources in the same area (triangulation across methods).
4.	Include individuals within the focus groups who are themselves directly involved in using and observing natural resources (thereby increasing the number of primary data providers).
5.	Use unequivocal categories of resource abundance.
6.	Ensure that the moderators of the focus group discussions have skill and experience in facilitating dialogue.

Development Goal 17 ("Strengthen the means of implementation and revitalize the global partnership for sustainable development") and the European Union (EU) principle of subsidiarity: that powers are exercised as close to the citizens as possible, in accordance with the proximity principle in Article 10(3) of the Treaty of the EU.

Below we present the experiences of three CBM programs in terms of demonstrating the quality of their knowledge products. In Greenland, the PISUNA program compared community members' perceptions with trained scientists' reports. The comparison of trends in abundance focused on 24 attributes that were summarized by 33 community members from 2009 to 2011 in Disko Bay. The community members and the professional scientists produced similar results for 12 attributes (Table 3.3). Only for two populations, nearshore Greenland halibut and breeding Arctic tern (*Sterna paradisaea*), was there disagreement between locals' and scientists' reports of trends in abundance. For 10 attributes, we were unable to locate any scientist-produced data in the published literature to allow for a comparison with the community members' findings.

The results suggest that this CBM program yields information that can be as reliable as that derived from professional scientist-executed monitoring.

In Alaska, the CBM program of the Yukon River Inter-Tribal Watershed Council (YRITWC) has created datasets that have been widely recognized for their high quality. YRITWC is an Indigenous nonprofit organization consisting of 73 Canadian First Nations and Alaska Native Tribes within the Yukon River Watershed. They have the vision "to be able to drink water directly from the Yukon River." When the organization was formed in 1997, they dedicated themselves to several tenets, one of which was "to understanding the Yukon River Watershed by means of monitoring, measuring and researching, and to use this knowledge to clean, enhance and preserve life." Indigenous peoples living within the river's watershed were concerned about maintaining the health of the river, as it is essential to supporting their way of life.

In 2006, 31 Indigenous governments in Alaska and Canada, the US Geological Survey (USGS), and YRITWC formed a partnership to develop a monitoring program on the Yukon River and its tributaries to study water quality indicators and thus monitor the river's health (Indigenous Observation Network, ION). The aim of the data collection was for the communities to inform community water resource planning and regional decision-making, whereas the USGS's interest was to investigate climate change indicators and water quality impacts. To be successful, the data needed to be considered reliable by both the partner Indigenous community members and the scientific community.

Table 3.3. Comparison of community members' perceptions and trained scientists' assessments of trends in abundance of sea ice, two human activities, and 21 populations of fish, mammals, and birds in northwest Greenland 2009–2011 under the Piniakkanik Sumiiffinni Nalunaarsuineq (PISUNA) program.

	Attribute	Perceptions*	Scientists' assessments	Source*	Correspondence
Fish	Atlantic cod, D	‡	Few data	Siegstad 2011	n.a.
	Wolffish spp., D	↑	↑ / ↔	Siegstad 2012	(✓)
	Greenland halibut	↑	↓ / ↔	Siegstad 2011; 2012	⊘
Marine mammals	Ringed seal	↓	Few data	Boertmann 2007; Rosing-Asvid 2010	n.a.
	Harp seal, D	↑	↑	Department of Fisheries and Oceans 2010; Rosing-Asvid 2010	✓
	Narwhal	‡	Few data	North Atlantic Marine Mammal Commission 2012	n.a.
	Humpback whale	↑	↑	Heide-Jørgensen et al. 2011	(✓)
	Minke whale, D	↑	↑	Heide-Jørgensen et al. 2010	(✓)
	Minke whale, U	↔	Few data	No information	n.a.
Land mammals	Arctic fox, D	↑	Few data	Boertmann 2007	n.a.
	Caribou, N	↔	↔	Cuyler et al. 2005; Cuyler & Nymand 2011	✓
	Musk ox, L	‡	Few data	No information	n.a.
Birds	Snow goose, D	↑	↑	Boertman 2007	✓
	Greenland white-fronted goose, U	↓	↓	Boertmann 2007; Boyd & Fox 2008	✓
	Canada goose	↑	↑	Bennike 1990; Fox et al. 1996; Boertman 2007	✓
	Common eider	↑	↑	Chaulk et al. 2005; Merkel 2010	(✓)
	White-tailed eagle, D	↑	Few data	No information	n.a.
	Large gulls,* D	↑	Few data	Boertmann 2007	n.a.
	Arctic tern, D	↑	↔	Boertmann 2007; Egevang & Frederiksen 2011	⊘
	Brünnich's guillemot, breeding	↓	↓	Burnham et al. 2005; Labansen & Merkel 2012	✓
	Little auk, D	↑	Few data	Egevang & Boertmann 2001; Boertmann 2007	n.a.
Other	Winter sea ice,* U	↓	↓	Danish Meteorological Institute	✓
	Offshore ships, U	↑	↑	Arctic Marine Shipping Assessment 2009	(✓)
	Trawling, D	↑	Few data	No information	n.a.

Sampling protocols were developed in partnership to ensure high-quality data collection. Over the years, ION has created a 10-year water quality baseline dataset from which to measure the health of the river. Over 300 community members have been trained in USGS water-monitoring protocols, and this has resulted in more than 1,500 samples collected at 54 sites along the 3,700 km length of the river, from the headwaters to the Bering Sea (Mutter and Fidel 2018).

These efforts generated a baseline record (long term at some sites) of water quality in the river basin, important for understanding climate change impacts. In interviews with Indigenous partners, data generated by ION was recognized as more credible than similar data collected by industry or government that did not incorporate input from residents. It was stated that this was because "our people" were the ones collecting the data, which were also found to be useful for decision-making at both regional and community level (Wilson 2017). These data have been used in 17 peer-reviewed scientific publications to better understand large-scale environmental and climate-associated changes occurring within the watershed.

In Canada, the Inuvialuit Settlement Region Community-Based Monitoring Program (CBMP) has established an elaborate plan for verifying their data and information. This program compiles bird, fish, and mammal harvest information each month from six communities to support decision-making by Inuvialuit organizations and co-management boards. Nine community resident technicians work together with local hunters and trappers committees, using iPads and an associated application to remotely upload data onto the Inuvialuit Settlement Region Platform. The program employs a four-step verification process, which involves (1) the community resource technicians comparing written data records with what has been uploaded to the online platform; (2) the hunters and trappers committees (HTC) comparing data to previous years, using their unique local knowledge to assess reporting

---

Legend: ↑, increased abundance; ↓, declining abundance; ↔, no major change in abundance; ‡, increased abundance reported in some areas, decline in other areas; Few data, little or no abundance data available; ✓, correspondence between community members' and scientists' assessments; (✓), probable correspondence between community members' and scientists' assessments but the time, area and/or temporal/spatial scale of the assessments do not match; n.a., not applicable; ⊘, no correspondence. D, Disko Bay; L, Naternaq/Lersletten and Svartenhuk; N, Nassuttooq/Nordre Strømfjord; U, Uummannaq Fjord. *For literature cited, Latin names and details, see Danielsen et al. (2014c).

Community monitoring of Yukon River has improved the understanding of climate-associated changes occurring within the watershed. It has also contributed to a large number of scientific publications. Credit: Edda Mutter

accuracy, and searching for anomalies and observations that are beyond the normal or expected range; (3) the resource person associated with each hunters and trappers committee verifying the paper records with the online version to check for data transfer errors; and (4) the CBMP coordinator verifying paper data records with the online platform and transferring relevant information to appropriate groups and/or managing anomalies and observations as observed by the HTC (Québec Workshop).

Conclusions: Ensuring the Quality of Knowledge Products

CBM programs are taking a number of measures to ensure the quality of the knowledge products they generate. Key potential measures include careful planning, explicit consideration of likely bias, and storing the data in its most disaggregated form and with details of exactly how it was collected. Other measures include carrying out checks to keep errors in recording and data storage at an acceptable level, and incorporating triangulation of the recorded data, or allocating individuals with rank based on their knowledge and allowing data to be disaggregated according to this ranking. Finally, as in any initiative, thorough guidance and supervision of the participants is important. As CBM programs differ in their goals and focus, some may not take these measures but can still be relevant and important.

3.7   Good Practice: Addressing the Rights of Indigenous and Local Communities

Many CBM programs offer successful examples of the use of Indigenous and local knowledge for sustainable environmental monitoring. In addition, many Indigenous communities are interested in implementing CBM activities because they see a direct link between documentation of environmental change and land and resource use and the capacity of communities to assert their rights (Zhigansk Workshop). Rights may relate to (1) land and resources and (2) the knowledge and information that belongs to Indigenous and local communities.

With regard to rights related to land and resources, many Arctic CBM programs have the primary aim of enabling Indigenous and local communities to be "heard" by management agencies, thereby promoting their rights to land and resources. One example is the CBM program of the new self-rule Thcho Government in Canada. This program has engaged Indigenous and local community members to monitor the Marian watershed, where mining is planned. Information from the observers has led to the relocation of planned paved roads to avoid fish habitats and the migratory routes of moose (Québec Workshop). Stories like this need to be told and broadly disseminated (see also section 4.5).

Other CBM programs that enable Indigenous and local communities to be "heard" by management agencies include those established by the Centre for Support to Indigenous Peoples of the North in Russia and the programs by SnowChange Cooperative in Finland, PISUNA in Greenland, and the Arctic Eider Society in Canada (further discussed in sections 3.1, 3.2, and 3.4).

With regard to the knowledge and information that belong to Indigenous and local communities, all CBM programs must follow principles of free, prior, and informed consent, and take care to protect the intellectual property rights of Indigenous and local communities.

Unfortunately, visitors do not always respect the rights of Indigenous and local communities. There are, however, protocols and procedures for enabling a respectful use of Indigenous and local knowledge. YRITWC, for instance, has published a sample agreement between researchers and Indigenous peoples that describes the expectations that Indigenous peoples may have of researchers when collaborating on projects (see docs.wixstatic.com/ugd/dcbdaf_0c50d62580124f2b9789ad6d1194d536.pdf).

Clarifying data ownership and data use rights is essential for all CBM programs. The Arctic Borderlands Ecological Knowledge Society (ABES) has been documenting local experiences of ecological change in the area of the Porcupine caribou herd since 1993. Local experiences are different from Indigenous knowledge, however, and many community members have pointed out to ABES over the years that most of the experts who are interviewed are Indigenous knowledge holders, and so the information they give is informed by Indigenous knowledge (Fairbanks Workshop). It is important that the data remains the property of those who create it. It is likewise important that the process for gaining access to the data is satisfactory to the communities involved. Currently, anyone wanting access to the data is required to apply to the communities they are requesting the data from. This process works to preserve and protect intellectual property but can be seen, incorrectly, as limiting access (Ashthorn 2018).

The Alaska Arctic Observatory and Knowledge Hub has developed an approach of respecting the data ownership of the Indigenous and local communities while at the same time making some data publicly available. The program contributes local observations to a database on sea ice, wildlife, and weather established by a predecessor project, the Seasonal Ice Zone Observing Network (Eicken et al. 2014), with the Exchange for Local Observations and Knowledge of the Arctic (ELOKA). The data come from daily narrative observations by sea ice experts and Indigenous subsistence hunters in Arctic coastal Alaska. The platform includes over 6,000 observations from 2006 to the present. Eleven coastal communities contribute to the database, although not all communities provide regular observations. There are two levels of access to the data: guest and registered user. The guest level of access contains summary information categorizing ice, weather, and wildlife observations but does not contain the full text of observations, which may contain sensitive information. Registered user access is provided to community members who have contributed observations. The database can be searched by hunter, community, data, and keyword. The information is generously shared with the public by the observers and the communities within which the observers reside. Before browsing or using the information in the database, however, guests must agree to adhere to a set of ethical and appropriate use guidelines, and to cite the data if it is used in publications (https://eloka-arctic.org/sizonet).

Making a living from the use of natural resources in the Arctic requires profound knowledge of the environment. Communities are often engaged in CBM to document environmental change and at the same time assert their rights. Disko Bay, Greenland. Credit: Michael K. Poulsen

It is important to provide further guidance to community members, management agency staff, and scientists in CBM tools and how to respectfully

Ensuring that decision-making processes not only use CBM data but also respect the rights of resource-dependent communities to shape decision-making is critical to effective CBM. Nenet Autonomous Okrug. Credit: Martin Enghoff

connect information across knowledge systems. The Arctic and Earth Signs program aims to provide a venue for CBM that encourages partnerships across multiple generations and knowledge systems. The project, based at the University of Alaska Fairbanks's International Arctic Research Center, is unique in that children (ages 5–14) and youths (ages 15–24) share a respected voice in co-identifying an issue to focus environmental monitoring on, along with a community team of educators, community members, elders, and scientists. After the team has received training on how to facilitate the process, it identifies an issue important to the community to be able to plan for rapid changes. The youth document Indigenous and local knowledge from elders and longtime community members, and deepen their knowledge of the topic through culturally responsive curricula. They co-design an investigation or monitoring project with their community team, using the options from a diverse array of monitoring protocols in the Global Learning and Observations to Benefit the Environment program (GLOBE; www.globe.gov). The youth are empowered to use the data to lead an environmental action or stewardship project that will help address the issue in their community, and to share their data and project with their Indigenous and local community as well as scientists and the international GLOBE community.

The Imalirijiit program in Canada is also training youth in CBM tools and how to respectfully connect information across knowledge systems. This program, which began in 2018, is beginning to collect baseline data on water quality and contaminants in local country food in the George River area in Nunavik prior to the opening of a rare earth mine. The program also develops the capacity of local youth in environmental science and interactive mapping.

Conclusions: Addressing the Rights of Indigenous and Local Communities

The Arctic CBM programs provide many examples of where the rights of Indigenous and local communities to land/resources and to protect their knowledge are being successfully addressed. Such experiences should be further disseminated. There are existing protocols for enabling a respectful use of Indigenous and local knowledge that should be made more broadly available. Data ownership and data use rights in CBM programs must be clear and follow principles of free, prior and informed consent. Further guidance should be undertaken in CBM tools and in how to properly address the rights of Indigenous and local communities.

3.8 Summary: Good Practice

In this chapter, we discussed good practice in Arctic CBM programs, based on a literature review and workshops with CBM programs in Alaska, Canada, Greenland, Norway, and Russia. When establishing CBM programs, representatives of community members should play a central role. The monitoring should reflect the priorities of the local communities and be kept as simple and locally appropriate as possible.

When implementing CBM programs, it is vital to have organizational and support structures that sustain the CBM effort from the community up to the management authority level. In recent years, the use of Indigenous and local knowledge for informing decision-making has received increased attention at policy level in the Arctic. Management authorities must enact policy in practice, listen to Indigenous and local community members, and provide feedback to the communities on how CBM results are used. Knowledge management platforms and other new data-sharing approaches may hold great potential.

To sustain CBM programs, they must ensure that the participants' observations are used for decision-making and that the participants are informed of how the information is being used. When there is high turnover of management authority staff, the negative effects may be minimized by involving multiple staff members in the CBM program. Greater impacts may sometimes be obtained through CBM by documenting the management interventions that result from CBM programs. Likewise, impacts may be obtained by bringing communities together, sharing information, and promoting advocacy on the importance of CBM-derived information.

With regard to connecting and cross-weaving with other approaches, some programs are intertwined with scientist programs at the method level, whereas others run independently and in parallel with scientist programs where these are available. At present, few CBM-derived datasets are included in Arctic data repositories. Representatives of Indigenous and local communities should decide whether data from their CBM programs needs to be linked with global repositories; the process must take place in accordance with their free, prior, and informed consent. When appropriate, CBM programs could connect with global repositories such as ELOKA.

CBM programs concerned with the credibility of their knowledge products can ensure careful planning, explicit consideration of likely bias, and storing of the data in its most disaggregated form, with details of exactly how it was collected. Other measures include carrying out checks to keep errors in recording and data storage at an acceptable level. As CBM programs differ in their goals and focus, some programs may not take these measures but can still be relevant and meaningful. Scientists could allocate some time and effort to visiting CBM programs and learning about them.

Addressing the rights of Indigenous and local communities is important to most CBM programs. The Arctic CBM programs provide multiple examples of where the rights of Indigenous and local communities to land/resources and to protect their knowledge are being successfully addressed. One critical measure related to the data derived from CBM programs is that data ownership and data use rights must be clear and follow principles of free, prior, and informed consent.

■ 4 Challenges and How to Address Them

This chapter examines a range of challenges encountered in implementing and conducting CBM in the Arctic, including challenges associated with the linkage between different CBM programs, between CBM and management authorities, and between CBM and scientist-executed environmental monitoring programs. For six commonly occurring types of challenge, we consider the extent, effects, causes, and potential interventions.

Part-time hunter in Uummannaq skinning a seal, an everyday activity in most communities in Greenland. Credit: Michael K. Poulsen

This chapter is intended to provide an overview of potential hurdles that CBM program facilitators, community members, collaborating scientists, or decision makers might face during the implementation and operation of CBM activities. The set of entries below is neither comprehensive nor ordered according to priority. Rather, it aims to inform the development of road maps toward knowledge co-production and resource co-management for new CBM efforts, and may serve as a reference for discussions in CBM programs already under way.

Like Chapter 3, this chapter is based on a review of the scientific literature and workshops with CBM program practitioners and community members engaged in CBM programs (see Table 1.1 for an overview of workshops).

4.1 Challenge: Limited Ability or Interest of Management Agencies to Access, Understand, and Act on CBM-Derived Guidance

Governments across various jurisdictional scales have demonstrated increasing interest in supporting community-based observation and monitoring in the Arctic and in identifying mechanisms for these programs to inform action. At the same time, CBM practitioners report continued challenges in working with governments to operationalize or act upon community observations in decision-making.

Extent

Government agencies and institutions reaching into the lives of Arctic communities and individual people may not be responsive to or act upon knowledge and information emerging from CBM activities. This is also the case with scientific organizations advising government agencies and institutions. This challenge can take one of several forms, including an unwillingness to engage with communities; an inability to access or ingest information from CBM efforts; or a lack of management follow-through on insights gained from CBM (Eicken 2010; Johnson et al. 2015, 2016).

By contrast, when there are specific policies that mandate direct communication between CBM programs and institutional actors for the purpose of regulating resource use, uptake of CBM information by agencies can be relatively straightforward. An example is the Eskimo Whaling Commission in Alaska, which provides harvest data directly to the North Slope Borough's Department of Wildlife Management and the International Whaling Commission to help set quotas. This direct communication channel was established through the Marine Mammal Protection Act (Albert 2000).

Causes

The range of causes underlying these challenges is broad. Despite recent progress (Armitage et al. 2011; Kendall et al. 2017), both government agencies and academia continue to struggle to understand the nature and relevance of CBM, and the Indigenous and local knowledge that informs many CBM efforts. Misconceptions include a perceived lack of reliability of observations derived from CBM and failure to appreciate equivalency of information generated through CBM with observations from scientists—the latter demonstrated by a growing body of literature (Johnson et al. 2015; Danielsen et al. in review). Power dynamics, history of governance and land or resource use rights, and conflicting viewpoints regarding present-day land and resource management issues may create an adversarial relationship between management authorities and community members (Fairbanks and Québec Workshops; Armitage et al. 2011; Eicken et al. 2011; Nadasdy 1999). Government staff tend to understand realities differently from community members. Moreover, involving community members in observational programs is only one of many factors in the policy-economy area that government staff need to consider. Further, CBM programs sometimes communicate data as a mixture of political viewpoints and evidence, involving advocacy-based evidence rather than evidence-based advocacy.

At the same time, a number of factors make it difficult for government agencies to respond proactively and rely on CBM as an important tool in responding to rapid Arctic change. These include bureaucratic and political hurdles to implement innovative approaches, such as CBM in a management context; potential disconnects between the legal or regulatory system and agency research divisions; and a lack of resources or expertise to implement good guidance on CBM implementation.

With respect to the last factor, not involving community members in designing observation programs is problematic. Natural scientists are typically involved in establishing CBM programs, but often lack an understanding of the broader dynamics of governance of land and resources. Moreover, while changing, the current reward system for scientists is still skewed toward peer-reviewed publishing and implementing the scientific process, rather than helping address real-world problems in a co-production of knowledge setting.

Finally, the international scientific bodies responsible for advising Arctic governments on resource management (and many other government agencies) are only now beginning to consider the establishment of procedures to take the observations and knowledge of community members into account for their advisory services (Alaska Arctic Policy Commission 2015; Nordic Council of Ministers 2015; Protection of the Arctic Marine Environment 2017).

Effects

Government agencies' inability or unwillingness to access and respond to community members' findings in their decision-making presents a major, decisive hurdle to the successful implementation and operation of CBMs. In the long term, it likely results in termination of CBM efforts. More important, it deprives both the communities and government from realizing the benefits of CBM-informed management of resources or response to Arctic change, such as opportunities for increased income and food security from sustainable resource use in small rural communities, or more effective response to community-level hazards associated with rapid Arctic environmental change. Decisions that impact communities, and don't take into account information from community-driven CBM, often lose credibility with community members.

Possible Interventions

- Make it easy for government agencies to access and use CBM data, and establish specific policies demanding government use of CBM data.
- Further develop best practices, protocols, and procedures to enable government agencies and international scientific organizations to incorporate local and CBM-derived information on natural resources and resource use in their decision-making (see also section 3.4).
- Bring communities together, share information, and promote advocacy on the importance of using information from CBM programs.
- Raise awareness (through meetings, publications, audiovisuals, etc.) within government agencies and international scientific organizations and management bodies on the value of local knowledge and observations.
- Develop two-way, direct government-community communication channels for sharing CBM information; ideally, these should be linked to specific mechanisms for making use of the data.
- Develop capacity among government staff and CBM leadership, and incorporate training in CBM into the curricula of the training institutions in the Arctic as piloted by the UArctic Thematic

Local observer Billy Adams tracks water characteristics in Utqiagvik, Alaska. Credit: Joshua Jones

Network on Collaborative Resource Management (www.uarctic.org/organization/thematic-networks/collaborative-resource-management/).
- Provide education and training on CBM activities as part of a research and monitoring portfolio, including training in collection of evidence from CBM programs.
- Develop monitoring and evaluation protocols for CBM programs that prioritize community feedback and involvement.
- Involve representatives of community members and CBM programs in the planning and evaluation of observation programs.
- Emphasize use of the CBM outcomes to government agencies and scientific organizations to show value of community investment in the development and sustained use of CBM programs (see also section 3.6).
- Emphasize community engagement in academic assessment and promotion.

4.2 Challenge: Insufficient Linkages Between CBM Programs and the Priorities of Northern Communities

While priorities vary across localities and regions, many Arctic communities prioritize individual and community health, economic opportunities (including improved employment opportunities), and other aspects of fate control, such as participation in the regulatory process or place-based education. In contrast, when introduced by individuals from outside the community, many CBM programs address topics that either fail to address community priorities or address them only marginally.

Extent

While CBM programs vary in their degree of community input into monitoring goals and attributes, there are also a number of CBM programs that respond directly to community information needs, including those that contribute to better-informed decisions or better-documented processes within fisheries, hunting/herding, transport/shipping, forestry, mineral and hydrocarbon extraction, and tourism (Chapter 2). These are key economic sectors in the Arctic. Community members are often motivated to participate in CBM programs to address critical socioeconomic and cultural issues for Arctic people such as "protecting rights to land and resources" and "sustaining the health and abundance of wildlife." Only in a small number of CBM programs do community members participate for financial benefit alone.

Causes

CBM program activities do not always align with the community priorities due to insufficient involvement of community members in their design, or due to assumptions made by nonresidents about the priority issues of Northern communities. Thus, university researchers often focus on pan-Arctic large-scale processes that may be of little interest at the local level. Government agencies may be constrained by legislation or regulatory frameworks on the type and scale of information that is collected.

Communities are diverse, and it can be difficult to identify and derive priorities for monitoring that reflect consensus (see discussion in Wheeler et al. 2016). Additionally, some CBM programs may not reflect the immediate priorities of community members but nevertheless may address issues of relevance to livelihoods of Northern communities, such as those that contribute to a better understanding of a particular component of the environment. When

Food on the table—mostly from natural ecosystems. Inside a Nenet chum, Indiga, Nenets Autonomous Okrug. Credit: Martin Enghoff

community members are robustly engaged in designing CBM programs, the result should be a focus that reflects at least some of the community's priorities for observing and monitoring.

Effects

If CBM programs do not address community priorities, community members are unlikely to contribute their time and resources to the observing efforts. In these programs, community members are typically only involved in the data collection phase (rarely in the design or interpretation). Since only one party benefits (the scientists or associated government agency), such programs are unlikely to be sustained over time.

Possible Interventions

- Involve representatives of community members and CBM programs in the planning and evaluation of observation programs (further discussed in section 3.1).

- Rely on a broader knowledge co-production and resource co-management framework to guide CBM implementation and use of CBM-derived information (Inuit Circumpolar Council—Alaska 2015; 2020, Lovecraft et al. n.d.).

4.3   Challenge: Sustaining Community Members' Long-Term Commitment to CBM Efforts

Beyond the challenges identified under 4.1 and 4.2, maintaining long-term interest and involvement by community members, and thereby ensuring continuity of CBM activities, can be challenging. Frequent staff turnover at the management authority level is a related problem (discussed in section 3.3).

Extent

Fatigue among community members and participant turnover at the community level were considered significant challenges for approximately one in five of Arctic CBM programs surveyed (Chapter 2).

Causes

One major reason for a loss of motivation and engagement among community members is a poor fit between the design of the CBM program and the local context, in particular community interests and concerns. Further problems can result from observation protocols that consume too much time or resources. Insufficient feedback on CBM results and outcomes also contributes to participant fatigue and disengagement. When designing CBM programs, it is also important to be aware that not all community members and communities are equally interested in and capable of participating. For instance, CBM programs aimed at protecting rights to land and resources and contributing to sustaining the health and abundance of wildlife tend to have greater sustainability when community members are heavily dependent on living resources for their livelihoods, with strong impacts on culture. In other communities, it may not be meaningful to establish such programs. Finally, it is important that all parties involved in a CBM program feel their effort amount to something, and that their contributions are recognized, including use of CBM information for actual management decisions at higher levels. Intellectual incentives are not sufficient.

Evenk community groups have documented that fish (Siberian and Arctic cisco, *Coregonus sardinella*, *C. autumnalis*) are swimming deeper in the Lena River due to warmer waters. They are therefore difficult to catch with the permitted net types. This observation has been used by the Republic Indigenous Peoples' Organization of Sakha Republic to influence changes in permitted net types. Yakutia, Russia. Credit: Martin Enghoff

Effects

A loss of motivation among community members and other hurdles to participation at the community level may lead to rapid and frequent turnover of CBM observers and contributors. Such turnover potentially jeopardizes the long-term sustainability of the CBM program, including the continuity of the resulting data records (Conrad and Hilchey 2011).

Possible Interventions

- Involve community representatives in planning and evaluating observation programs (see also section 3.1).
- Use tools and approaches for data collection that can easily be incorporated into the day-to-day activities of the community members and that allow the results to be incorporated into decision-making.

- Provide regular feedback to community members with the findings and results of the CBM programs and examples of how these findings are being used for decision-making (see section 3.2).
- Motivate all parties in the CBM programs, from community members to the authorities involved. Consider carefully what each party gains from participation. For many participants, there is a strong incentive if the concept of their voice in society has changed as a result of the effort.
- Different incentives may need different methods of communication, varying from social media and gamification (employing game design elements) to more tangible benefits such as caps, rubber boots, or financial compensation.

4.4 Challenge: Lack of Compatibility Between Data Formats of Scientist-Executed Monitoring and CBM Programs

In CBM programs, where community members and facilitators are keen to connect with scientist-executed programs, lack of mutual visibility and compatibility between the two kinds of datasets generated is a major hurdle.

Extent

The incompatibility of data formats among scientist-executed and CBM programs in the Arctic is reported as a challenge, but the extent of this problem is poorly explored (e.g., Pulsifer et al. 2012). Furthermore, not all CBM programs intend to connect with scientist-executed programs (Fidel et al. 2017), and vice versa, and for such programs, differences in data formats may not be a concern. However, incompatibility of data formats may prove problematic down the line if program goals change, and scientists or community members determine that there are benefits to shared data. Additionally, even when there is interest in making data compatible, CBM programs and scientist-executed programs are not always visible to one another; it can be difficult to identify programs with relevant data operating at very different observing scales (section 3.5).

Causes

Scientist-executed and CBM programs are usually developed independently. They are based on different realities and epistemologies or worldviews (Tengö et al. in review). They may be connected, yet not directly comparable. They often have different aims and use different tools and approaches. For example, CBM programs are typically user-driven with a local focus and a single intended application. In contrast, global and regional observing programs tend to feed into data management programs that focus on integrating data that arise from multiple sources holistically, such that they can be used in a range of applications.

There is clear value in incorporating CBM programs into global and regional data programs, given that measurements tend to be useful to multiple applications, but there are several obstacles to doing so:

- Relevant global and regional data archival facilities are not visible or accessible to CBM programs and vice versa.
- Format and modality of data entry, transmission, archival, and related restrictions enforced by regional and global data centers exclude CBM contributions.
- Intellectual property rights, licensing, acknowledgment, and citation concerns.

Scientists incorporating CBM programs and tools into their work have found that their science has become better (e.g., Eerkes-Medrano et al. 2017; Mercer et al. 2010). It is particularly important that CBM programs store data in their most disaggregated form and with details of how they were collected, and that the raw data in the CBM program are kept for checking and reinterpretation (Québec Workshop).

Effects

When the data formats of scientist-executed and CBM programs (including international scientific data repositories) are incompatible, it may be difficult or impossible to connect the programs at the data level. Even when they are compatible, mutual limitations in knowledge mean that synergies go unrealized. Opportunities for obtaining a larger-scale understanding of environmental questions are thus lost.

Possible Interventions

- Encourage managers of scientific data repositories to adjust their data formats so they become receptive to data from CBM programs.
- Highlight examples of the successful incorporation of both CBM and scientists' program data into publications and assessments to encourage further cooperation.

- Undertake awareness raising (meetings, outreach, education) with international scientific organizations on the usefulness of incorporating information based on Indigenous and local peoples' knowledge into scientific data repositories in order to obtain a better basis for future decision-making.
- Identify CBM programs that are keen to connect with international scientific data repositories and which are not currently connected, and encourage them to identify suitable scientific data repository partners and develop further cooperation (see section 3.5).

4.5 Challenge: Intellectual Property Rights, Respect and Reciprocity, and Free, Prior, and Informed Consent (FPIC)

Respecting the rights of participating Indigenous and local communities is an important aspect of all CBM programs. Additionally, those that engage Indigenous knowledge must be rooted in an awareness of sensitivities specific to the management of data related to Indigenous knowledge. This will have an impact on how the programs are implemented.

Extent

Many CBM programs collect sensitive information, such as harvest locations of particular species (which hunters may not want to share), or about sacred or sensitive sites that require special protection. Most programs work carefully with community members to identify and address these sensitivities. In some cases, disrespect for intellectual property rights and proper consultation has been reported (Fidel et al. 2017); however, little is known as to the actual extent of this challenge. Many communities across the Arctic have concerns about ethical lapses and issues related to data management in past projects that have been implemented on or near their lands and territories, and these concerns shape their receptiveness to proposed collaborations in the present (Gearheard and Shirley 2007).

Causes

CBM programs operate within a broader context of research practice in which communities are often approached by well-intentioned outsiders interested in collaboration but without a long-term commitment to understanding the local context of knowledge production and use. Many Arctic Indigenous community members can name multiple instances in which they have shared their knowledge with visiting researchers and not received any tangible product or benefit from the collaboration. As a result, communities are increasingly asking for greater awareness and sensitivity to ethics in research practice from the Arctic research community at large (Fairbanks and Québec Workshops). This includes the need for awareness of and respect for existing protocols and frameworks for meaningful engagement of Indigenous peoples based on Indigenous rights, such as free, prior, and informed consent (FPIC). Some guidelines, such as the Tkarihwaié:ri Code, describe how to promote the use of Indigenous and local knowledge (Convention on Biological Diversity 2011). However, the process described in the guidelines is very time-consuming and difficult to implement in practice. Indigenous communities and organizations have raised the need for regionally appropriate and specific ethics protocols and research agreements (Nickels and Knotsch 2011).

Some research and CBM programs have unclear agreements on data ownership and data use. It is important that the communities maintain control over data, that data is accessible to community members, and that a long-term data storage solution is identified as part of CBM program design. If the community wants to delegate data management to other organizations, then that should be their choice.

CBM can be an important step in Indigenous and local communities' efforts to claim their rights, for example, by framing land rights in relation to use and occupancy (Zhigansk Workshop), as well as their right to knowledge and their share of any benefits accruing from this knowledge through, for example, the access and benefit sharing mechanism of the CBD. This, however, requires that FPIC is fulfilled and that there are clear agreements on data ownership and data use.

Effects

In order to secure the full engagement of community members, CBM programs must demonstrate awareness of and sensitivity to the importance of protecting intellectual property rights and sensitive knowledge. Setting up CBM programs without the consent of the Indigenous and local communities means that community members are unlikely to be happy with the results, and the programs will not be sustained. In the worst case, programs set up without local consent may exacerbate conflicts between government agencies, scientists, and community members.

CBM programs are occasionally criticized for making data available for "mining" by outsiders. Internationally, there are a number of examples of the misuse of Indigenous and local knowledge, including private companies using

There are many steps ahead before Arctic CBM programs become organizationally and financially sustained. Local observers in the boreal forests in Komi, Russia. Credit: Martin Enghoff

this knowledge for their own benefit without providing any compensation. For example, a 1988 *National Geographic* article that described the *tiki uba* plant used as an anticoagulant by the Amazonian Ureu-Wau-Wau tribe attracted the attention of researchers for the pharmaceutical company Merck. After successful testing, Merck commercialized the product, useful in heart surgery. The company used Indigenous and local knowledge without having any obligations to compensate the Ureu-Wau-Wau tribe (McIntyre 1989; Posey 1998).

Possible Interventions

- Highlight examples of the successful use of Indigenous and local knowledge in sustainable CBMs.
- Further develop suitable protocols and procedures for enabling a respectful use of Indigenous and local knowledge.
- Clarify data ownership and data use rights (further discussed in section 3.7).
- Carry out further training in CBM tools.

- Reward CBM programs in which scientists and community members work together to develop innovative data management tools that address community concerns.
- Encourage funding agencies to make ethical data management a central component of proposal review.

4.6 Challenge: Organizational and Support Structures for CBM Programs

The success of CBM programs and activities hinges on organizational and support structures that sustain the effort from the community level up to the government level.

Extent

Insufficient organizational and support structures (roles and responsibilities, communication lines, funding, etc.) are recurrent challenges for some CBM programs in the Arctic.

Causes

Several factors contribute to this problem. Sometimes, programs are established without any insight into the existing organizational or institutional landscape in the area. Programs may therefore establish parallel "island" structures instead of properly incorporating CBM activities into existing organizations. Like all observing programs, CBM programs must deal with a mismatch between the short-term nature of funding cycles and the importance of sustained data collection. This can limit their ability to institutionalize CBM activities within existing organizations. CBM programs are often more successful when they involve a strong local organization as a coordinating entity; a lack of a strong local institutional partner can present a barrier to sustaining programs over time (Québec Workshop). Similarly, for programs aimed at informing resource management or land use planning (or other types of decision-making), involving relevant decision-making entities in program planning is critical to uptake of information in the long run (Québec Workshop).

Effects

Insufficient organization and support can result in CBM programs being short-lived and unable to attain their objectives. CBM programs may also be initially developed with scientists in academic institutions that provide organizational, administrative, or technological support, but then need to transition these support roles to an appropriate community-run entity over the long term. Organizations like local hunters and fishers groups, cultural heritage institutions, and women's groups provide essential information about community expertise and are able to coordinate logistics and other needs. CBM programs developed without sufficient involvement of local institutions (or without sufficient allocation of resources to support this involvement) risk creating instabilities in staffing and data management over the short and long term (Johnson et al. 2015). Similarly, programs interested in informing decision-making create weaknesses when they fail to engage personnel from relevant institutions, such as government agencies. This can be challenging, however, when staff turnover in these institutions is high and there is a lack of organizational memory or coordination to sustain collaborative involvement.

An example of a more coordinated approach to research with respect for the local community's time and resources is the Svalbard Social Science Initiative (www.svalbardsocialscience.com), which was launched in 2019 to coordinate social science research in Longyearbyen, Svalbard. Its aim is to create linkages among social scientists working with issues related to Svalbard, establish a platform for coordinating research, and facilitate communication with local communities and other scientists. The initiative has been appreciated by the local council and local actors, and has opened the door for more systematic meetings and dialogue about the challenges and possibilities for Longyearbyen and Svalbard, as well as how research can be of common interest and be undertaken in a collaborative and coordinated way.

Possible Interventions

- Include sustainability in the CBM program from the start. When developing a CBM program, understand and respect existing political and organizational structures in the area, so as to build on and not undermine them (further discussed in section 3.3).
- Develop international partnerships and funding structures to increase local support and collaboration options for CBM efforts.
- Be aware that institutionalizing CBM programs within existing organizations is a capacity-building process that takes time and must be based on trust and confidence.
- Consider local institutional capacity when selecting community partners prior to establishing CBM programs.
- Where multiple CBM programs exist in the same community, consider coordinating and pooling support for local involvement to increase funds available for local capacity and employment

(for example, a shared CBM program coordinator at a larger percentage time equivalent).
- Maintain strong communication channels (and invest the necessary time to communicate) with all program partners.
- Build sustainable funding structures to support CBM.

4.7 Summary: Challenges and How to Address Them

In this chapter, we discussed some of the common challenges that Arctic CBM programs face, their causes and effects, and possible interventions. Limitations in the ability or interest of management agencies to access, understand, and act on CBM-derived guidance can create challenges for sustaining the engagement of communities, who expect to see that their hard work will lead to some kind of action. These limitations are rooted in both misconceptions about the value of CBM and in logistical and bureaucratic barriers. Possible interventions include awareness-raising and capacity-building among agency staff, and developing policies and tools to make it easier for agencies to utilize community observations.

A related challenge is insufficient linkages between CBM programs and priorities of northern communities. This can arise when CBM programs are informed more by goals of scientific researchers or management agencies than by community members, and as a result, fail to support local priorities such as health and wellness, economic opportunities, and transmission of local or Indigenous knowledge and place-based skills. Interventions include increasing involvement of community members in planning and evaluating CBM programs, and utilizing a knowledge co-production framework to guide implementation.

Another challenge is that of sustaining community members' long-term commitment to CBM efforts, which is influenced by both community perceptions of relevance and reward and by factors related to community capacity. CBM programs that are designed with significant community input and that directly address community information needs are more likely to sustain interest over the long term. Programs in which observing protocols are too time-intensive or that provide insufficient feedback to communities about outcomes risk burnout over time. Possible interventions include incorporating data collection into routine activities, prioritizing communication and feedback to community members about uses of CBM data, and considering how to motivate all parties, including community members as well as authorities.

Lack of compatibility between data formats of scientist-executed monitoring and CBM programs presents a barrier to broader use of CBM data. Possible interventions include working with international scientific organizations and managers of scientific data repositories to better understand CBM data, make use of CBM data where available, and to adapt repositories so that they are better able to receive data from CBM programs.

An additional challenge lies in a lack of broad awareness within the scientific community of intellectual property rights, the need for respect and reciprocity in working with Arctic communities, and free, prior, and informed consent. This challenge has both historic and contemporary dimensions, and affects the interest, capacity, and long-term commitment of Arctic community members, scientists, and management agencies to participate in long-term collaborations. Some possible interventions include development of regional and local protocols that reflect research and monitoring priorities and practices supported by communities, clarifying data ownership and use rights, and rewarding CBM programs that develop innovative ways to address community data management needs.

A final challenge is insufficient organizational structures to support CBM programs over time. Possible interventions include a focus on capacity-building of local institutions, assessing local institutional capacity prior to establishing CBM programs, and pooling resources among CBM and other observational programs so that funds for capacity-building and local incentives such as job creation can be coordinated.

5 Moving Forward with CBM in the Arctic

Across the Arctic and around the world, interest in community-based monitoring is growing (Kutz and Tomaselli 2019; Pecl et al. 2017). This reflects the environmental and social challenges humanity faces (IPBES 2019; IPCC 2019). There is broad consensus that the challenges require new knowledge and new forms of action across multiple scales, from the individual household through communities, subnational regions to national and international bodies (Díaz et al. 2019). New forms of organization, such as those that directly link and mobilize insights between local communities in different parts of the world, are emerging (Connors et al. 2012). Within this context, CBM offers the potential to support more informed governance practices that could guide decision-making in support of both local and global goals.

Global governance frameworks are increasingly emphasizing equity and local stakeholder inclusion in decision-making, although what this means in practice remains contested (Adayeye, Hagerman, and Pelai 2019; Green 2005). International agreements and frameworks for cooperation, such as the European Charter on Participatory Democracy in Spatial Planning Processes, emphasize the need for government agencies and scientists to collaborate with local-level actors who are directly impacted by decisions connected to their place and the local environment. There is growing recognition that in order to "leave nobody behind" in efforts to fulfill the UN Sustainable Development Goals, co-creation and co-management with local actors and respect of the knowledge and experience of Indigenous and local communities will be crucial.

Youth listen to local observers presenting the results of community-based monitoring in Yakutia, Russia. Credit: Martin Enghoff

Increasing awareness of the important role of local expertise in global and regional response and management frameworks has prompted institutions and bodies funding scientific research to give more consideration to CBM. Thus, researchers are increasingly encouraged or required to demonstrate societal impact of research projects, in particular benefits accruing to local communities (Interagency Arctic Policy Research Committee 2018). They are also stressing the need for cross-disciplinary research, data management, and data integration. Some recent funding calls, such as the US National Science Foundation's Navigating the New Arctic, also emphasize the importance of Indigenous knowledge in convergent approaches to research.

While there is greater interest in and acceptance of the contributions of CBM to Arctic research and decision-making, gaps in understanding and acceptance remain (Wheeler et al. 2020). Scientists who are not familiar with CBM programs sometimes argue that because Arctic phenomena often encompass large areas (for instance, ocean current, or the range of migratory species), the community knowledge is not relevant for informing decision-making processes because individual community members only make observations in small areas. While community members do obtain their knowledge from the areas where they live and travel, they are also sometimes able to connect their observations and knowledge to describe environmental phenomena and processes that happen at larger scales (Eicken et al. in review). We have provided many examples of Arctic CBM programs that inform decision-making processes from the local to the national level. We hope these examples can help stimulate a more nuanced discussion.

Below, we offer a few concluding observations about areas where we believe the field of CBM would benefit from either more experience or more research as we move toward a larger role for CBM in Arctic observing. These insights are emergent and may be refined through further reflection and study in the future.

Where do we need more experience and greater proficiency?
Although CBM is a centuries-old practice (section 2.2), the systematic study of CBM is a relatively young discipline. There are a number of fields in the CBM landscape where there is need for more experience and greater proficiency in drawing on CBM for broader benefits.

First, there is a need for more experience among international management bodies in mobilizing CBM data to inform their guidance to Arctic governments (Hovelsrud and Winsnes 2006). Among the international management bodies of greatest importance to the lives and livelihoods of Arctic community members are NAMMCO (North Atlantic Marine Mammal Commission), CITES (Convention on International Trade in Endangered Species of Wild Fauna and Flora), NAFO (Northwest Atlantic Fisheries Organization), and ICES (International Council for the Exploration of the Sea). Whereas the international management bodies are supposed to incorporate Indigenous and local knowledge into their advice to governments, this rarely happens in practice (Inuit Circumpolar Council 2018; Nordic Council of Ministers 2015). This is one of the reasons why the decision-making by Arctic government agencies—for example, on setting quotas and resource management—still only consider Indigenous and local knowledge to a limited extent. Advances in online digital platforms make it possible to share community-produced observations across sites and scales of decision-making, but coordination is still in its early stages, and such tools are not being fully used by international management bodies (Johnson et al. in review).

Second, there is a need for more experience in extending existing CBM programs to promote decentralized governance of natural resources in the Arctic. For example, increased local management of nearshore Greenland halibut has been proposed in Greenland. One way of implementing this would be to allow the local authority to increase the halibut quota by, for example, 10% once a quarter if the local authority can document that the average size of individual halibut is over a certain size (Ministry of Fisheries, Hunting and Agriculture 2011). The central government would still have to set the quota every year, define an "ideal" size of halibut that can be legally caught, and establish tools for local communities and local government authorities to document the size of halibut caught. Likewise, there is limited understanding of the role of Greenland halibut in Arctic ecosystems. There is a need to co-develop and test with local communities tools that they could use for monitoring fish stocks as part of a decentralized nearshore halibut fishery, perhaps building on the experiences of the existing CBM program, PISUNA (see sections 3.4, 3.5, and 3.6).

Valuable experiences in promoting decentralized governance of Arctic resources might be obtained from the Ivittuut region in southern Greenland, where scientists and community members have developed a community-led muskox harvest calculator (Cuyler et al. 2020). The harvest calculator is based on a simple, user-friendly demographic model. Community members' own observations of muskox abundance, recruitment, and mortality, provided in the Greenlandic language, serve as input. The calculator can enable that community members independently of scientists can undertake multiyear planning of local muskox harvest to ensure both a continued supply of meat for subsistence and of old bulls for guided trophy hunting.

Third, there is a need for obtaining experiences in using CBM in innovative ways, such as for documenting environmentally and socially sustainable production as part of certification processes and eco labeling (Raynolds and Bennett 2015). Small rural communities engaged in CBM could sell their products at higher prices if the products are labeled such that consumers know that purchasing the product is contributing to sustainable development, environmental conservation, and securing the rights of marginalized producers. Aside from fish, these products could include meat, fur, down, handicrafts, and tourism commodities. CBM is a simple and low-cost yet transparent approach to monitoring. A proportion of the extra money earned by the small-scale producers could be used to finance local community observing and management work. In this way, community observing could add value to a product or commodity, and the community observing activities would also receive financial support from the producers, helping to sustain the monitoring (see also sections 3.3 and 4.3).

Fourth, there is a need for finding effective ways of enhancing the capacity of government staff in CBM. Upscaling, extending, and promoting CBM

Lapland rosebay (*Rhododendron lapponicum*) at Nuussuaq Peninsula, Greenland. Credit: Martin Enghoff

require that an increasing number of resource managers and scientists are able to facilitate, implement, and operationalize participatory approaches to natural resource management in practice. So far there are very limited experiences in in-service training of government staff in CBM techniques in the Arctic. A course program for government staff in Greenland implemented through the University of the Arctic could serve as a pilot model (UArctic 2020).

Fifth, there is a need for more experience in facilitating environmental monitoring in rarely visited places of the Arctic. One option is improved and more widespread environmental monitoring efforts on the part of expedition cruise ships (Wagner et al. n.d.). The large expanse of the Arctic and the many remote parts that are rarely visited present challenges to environmental monitoring. Expedition cruise ships are increasingly reaching otherwise rarely visited places. Tour guides and passengers can contribute meaningfully to environmental monitoring in the Arctic (Poulsen et al. 2019). The monitoring may include observations from guides and guests, photographs, or the taking of water, ice, or soil samples for later analysis by scientists. The receivers of the data, samples, and reports may include cruise guests, cruise guides, relevant databases, Indigenous organizations and research institutions, as well as the authorities responsible for recommending or deciding on management actions. Some cruise operators are already participating in environmental monitoring of wildlife and cultural sites. It may be possible to learn from existing efforts, build on these, and extend participatory monitoring to even more cruises, but there are few documented experiences to date.

Furthermore, the increase in Arctic ship-based tourism also creates new threats to cultural heritage sites and local residents' hunting and harvesting areas (Arctic Council 2009). From the community perspective, there is a need for greater proficiency in monitoring the activities of ships and tourists who enter their regions. Indigenous guardian programs that hire community members to serve as guardians of traditional territories are growing in popularity. Guardians conduct regular patrols of traditional harvest and usage areas and cultural heritage sites, share information with visitors about opportunities and limitations on access to certain areas and resources, and collect monitoring data to uphold Indigenous governance (Peachey 2015). Further experience and greater proficiency in developing such programs are much needed. Ideally, future tourism, whether involving ships or community infrastructure, would draw on both scientific and community-based monitoring approaches to serve as a way to reaffirm shared values between Indigenous and other residents and visitors to the region, as well as to assemble observations that are critical to learning and sustainable management of seascapes based on recognition of Indigenous sovereignty and rights.

Where do we need more research?
We have highlighted a number of topics centered on CBM where current knowledge is nonexistent or limited. Moving forward, it is worth considering where research into CBM would be desirable, and where Arctic CBM experiences are of value to other efforts across the globe. As interest in using CBM grows, the field of studying CBM and its scientific and societal contributions should also develop. A growing community of practice could help speed up progress in CBM research.

Being able to conduct workshops and interviews to help document the breadth of knowledge that links locally observed phenomena to the understanding of phenomena at larger scales would be a useful practice to demonstrate to other researchers and decision makers that local CBM observations are not simply locally relevant anecdotal information (Johnson et al. 2014). An example is the contribution of Indigenous knowledge from the Bering

Sea region of Alaska to the 2019 Arctic Report Card issued by the National Oceanic and Atmospheric Administration. These observations of change were documented in a workshop with elders and knowledge holders with the specific goal of informing the widely read and cited report card (Slats et al. 2019).

An important way for community members to connect their observations and reach decision makers is through online digital platforms for storing and transmitting data (review in Johnson et al. in review). We believe digital platforms for CBM programs merit more attention. The pace of development drives techonology prices down, leading to wider deployment and adoption at the community level. Although internet access remains a barrier for some programs, this is becoming less of a challenge. We anticipate that increased access, combined with improved literacy and capacity and an interest in data sovereignty, will create greater demand within communities for platforms that provide data in formats that are relevant to local and larger-scale decision-making and use.

Another example of the need and great potential for additional research involves studying the implementation of CBM for global climate change adaptation measures. In 2018, in Katowice, the world's governments agreed on a rulebook to implement the Paris Agreement on climate change. The rulebook emphasizes the constraints encountered by governments and the United Nations Framework Convention on Climate Change in measuring the impacts and effectiveness of climate adaptation measures, and invited academia to develop methodologies to monitor adaptation efforts (Ford et al. 2015). Since effective monitoring of adaptation measures requires knowledge of what is happening on the ground, and must be inclusive and accountable, the implementation of CBM seems to be a good fit to support such efforts. There is a need for a monitoring process that captures detailed, disaggregated information on the social and environmental adequacy and effectiveness of adaptation measures. Such information would need to be both location-specific and of high resolution, but also capable of contributing to the broader data collection efforts in support of national and global goals. Our experience suggests that CBM could make a major contribution to these efforts in the Arctic and beyond; however, robust research is needed to test and examine CBM along with other participatory methods of monitoring adaptation measures.

Our analysis of Arctic CBM presented a starting point for understanding the motivations for sustained engagement by nonresearch professionals in monitoring activities (section 2.2). Without sustained community support and engagement in CBM, there is little chance that a CBM effort will succeed. We contributed the findings from Chapter 2 to a broader discussion on advancing the understanding of citizen science through meta-analyses at the

Citizen Science Association Conference in North Carolina in March 2019. We learned that the research community involved in CBM, and more broadly in citizen science, has only just begun to understand what motivates people to remain involved in CBM over the long term.

Our study of community-based monitoring programs in the Arctic provided insight for understanding the scope of CBM capabilities, common challenges, and good practices. Although not all CBM programs are specifically designed to influence decision-making, it was important to highlight their relevance for natural resource management and international environmental agreements. CBM is an evolving practice and a valuable tool for Arctic observing responsive to community interests. We hope our study has added some rigor to a field that has had its share of token efforts or unsupported claims of community involvement.

In particular, we have tried to highlight good practices, challenges, and ways to address them, which we hope will be useful to practitioners who want to start CBM programs as well as those programs interested in assessing and strengthening impact. Awareness and understanding of the importance of support structures for sustaining CBM programs during the program design phase may assist with the selection of communities and partners. Similarly, understanding that communities require feedback and communication about how data from CBM programs are being used to inform decisions may prompt researchers and communities to build strong linkages with government agencies and other potential users from the outset. Ongoing awareness of issues related to ethical data management practices and the importance of community stewardship over data, including the principle of free, prior, and informed consent, is similarly critical to establishing successful CBM programs.

The issue of sustainability in CBM remains an ongoing challenge, with CBM practitioners frequently emphasizing the lack of sustainable funding mechanisms as the main challenge, as well as turnover in community participation and community fatigue. From the funding perspective, CBM programs are no different from other researcher-driven long-term monitoring projects where it takes a dedicated researcher (or research team) to consistently apply for funding at regular intervals. Recommendations for improved national and international funding mechanisms have been made in the past to help improve sustainability of funding for CBM (Johnson et al. 2016), but more important, there is an underlying need to build capacity within communities and collaborating institutions that would allow CBM programs to keep running in a cost-effective manner.

As recognition of the value of CBM grows among researchers and decision makers, there will be a need to balance outside interests with community values. We anticipate a growing community of practice developing for those involved in CBM. This, in turn, would support new research to better understand the motivations for community participation in CBM and improve our understanding of the interconnections between CBM, other scientific research, and resource management and decision-making. Such insight would greatly improve efforts to sustain CBM programs in the Arctic and beyond, where community involvement in environmental observing provides information and community engagement in resource management that would otherwise be difficult to obtain.

Appendix A. Arctic CBM Programs

Title	Link	Content
A-OK (Alaska Arctic Observatory and Knowledge Hub)	eloka-arctic.org/sizonet/	Access to all data via database search function
Arctic and Earth SIGNs	www.globe.gov/web/arctic-and-earth-signs	Program information
Arctic Borderlands	www.arcticborderlands.org	Program information about Arctic Borderlands Ecological Knowledge Society and link to guidelines for requesting access to data
Bird phenology	dataverse.no/dataverse/uit	Large amount of open research data, including bird phenology data
	dataverse.no/dataset.xhtml?persistentId=doi:10.18710/4MCRQL	Data on first arrival dates of spring migrants
BuSK (Building Shared Knowledge)	katersaatit.wordpress.com/dyr-og-fangst/	Program information in Greenlandic and Danish language
	asiaq.maps.arcgis.com/apps/View/index.html?appid=72a0b7241ef341b0af3fb3812eddb320	Access to data
CSIPN (Centre for Support to Indigenous Peoples of the North)	www.uarctic.org/member-profiles/russia/8440/centre-for-support-of-indigenous-peoples-of-the-north-russian-indigenous-training-centre	Program information in English language. Use www.csipn.ru for additional information in Russian language.
Evenk & Izhma Peoples	N/A	
Fávllis	site.uit.no/favllis/	Program information, as well as Indigenous and local knowledge, in Norwegian
	www.meron.no/nb/	Data from Fávllis are/will become available via this link.

Appendix A. Arctic CBM Programs

Title	Link	Content
FMI (Finnish Meteorological Institute) Snow Depth	globefinland.fi/glofin/projektit.html	Program information in Finnish, with link to data in English, as well as information on FMI snow depth measurement campaign
Fuglavernd	fuglavernd.is/verkefnin/gardfuglar/gardfuglahelgi/	Program information in Icelandic
George River	N/A	
Farmers and Herders	See "Summer Farmers and Small Herders"	
Hares	haran.fo	Harvest data including graphics, in Faroese
ION (Indigenous Observation Network) Yukon River	www.yritwc.org/science	Program information in English
	www.sciencebase.gov/catalog/item/573f3b8de4b04a3a6a24ae28	Water quality data example
Local Environmental Observer	www.leonetwork.org/en/docs/about/about	Program information in English
	www.leonetwork.org/en/	Access to data
Marion Watershed	N/A	
Nordland Ærfugl	www.eiderducks.no/?side=om-nordland-aerfugllag&language=no	Program information in Norwegian
Oral History	www.snowchange.org/efforts-in-the-skolt-sami-areas-of-naatamo-watershed-finland/collaborative-management-along-the-naatamo-watershed/	Program information in English
	www.snowchange.org/pages/wp-content/uploads/2015/09/Snowchange-Discussion-Paper-9.pdf	Example of available reports
Pilot Whale	heimabeiti.fo/hagtol	Harvest data from 1584 to 2020 in Faroese
Piniarneq	www.sullissivik.gl/Emner/Jagt_Fangst_og_Fiskeri/Jagtbevis/Fritidsjagtbevis_samlet?	Links to Piniarneq harvest reporting only (Greenlandic and Danish). No information about how to get data, but data is likely to be available upon request.

Appendix A. Arctic CBM Programs

Title	Link	Content
PISUNA	www.pisuna.org/uk_index.html	Program information in Greenlandic, Danish, and English; and community reports in the original language and format
	eloka-arctic.org/pisuna-net/en/	Searchable database
Renbruksplan	www.renbruksplan.se	Program information in Swedish and local language
	www.sametinget.se/116236	Information on database under construction in Swedish and local language
River Owners Iceland	www.angling.is/en/catch-statistics/	Link to data in English and Icelandic
Sea Ice for Walrus Outlook	www.arcus.org/siwo	Program information in English
Seal Monitoring	selasetur.is/en/research/557-2/	Program information in English and Icelandic
Summer Farmers and Small Herders	www.slu.se/en/Collaborative-Centres-and-Projects/swedish-biodiversity-centre1/Research/projects/ongoing-phd-projects-/	Some program information in English
Walrus Haulout Monitoring	N/A	
Wildlife Triangles	www.nrcresearchpress.com/doi/pdf/10.1139/cjfr-2015-0454	Program information in English
Wild North	rannsoknasetur.hi.is/university_icelands_research_center_husavik	General information on the research center
WinterBerry	sites.google.com/alaska.edu/winterberry/	Program information in English
Älgdata	www.algdata.se/Sv/Pages/default.aspx	Program information in Swedish
	www.algdata.se/Sv/statistik/Pages/default.aspx	Link to data in Swedish

Appendix B. CBM Practitioner Questionnaire

The CBM practitioners in our survey were asked general questions relevant for all Arctic monitoring systems, including both scientist-executed and community-based monitoring programs, and questions of particular relevance to CBM programs.

The general questions were about the respondent, the general characteristics of the attributes observed, the sustainability of the monitoring, and the use and management of the data (see Questionnaire A, available at https://intaros.nersc.no/node/651).

The questions of particular relevance to CBM programs are presented below.

Questionnaire: Community-Based Monitoring Program

Section I: Central Questions
Section II: General Information
Section III: Community Members
Section IV: The Data

I. CENTRAL QUESTIONS

1. **What is the aim of the monitoring program?**

   (Examples: To protect rights over land and resources; To ensure sustainable use of resources; To protect threatened biota; To obtain a better understanding of the environment; Monitoring is just part of "everyday life"; Other)

2. **Who do you consider to be the users of the data/results from the monitoring program?**

   Write who you believe/know make use of the data.

   Other community members
   Government agency
   Civil society organizations
   Academic institutions
   Do not know
   Other _____

3. **Does the monitoring program link to natural resource governance (management of the resources), or to scientific research?**

   Link to natural resource governance
   Link to scientific research
   No links to natural resource governance or scientific research
   Do not know
   Other _____

4. **Do you supply or pass on your monitoring data to other organizations?**

   Yes
   No
   No, but we would like to
   Do not know
   Other _____

5. **If yes, to whom?**

   Select all you agree with.

   Government agency
   Civil society organization/NGO
   Academic institution
   Other: _____

6. **Has the monitoring contributed to the local community—and how?**

   Positively. Development of pride or self-esteem (COGNITIVE EMPOWERMENT)

   Positively. Participation in decision-making, increased local governance (POLITICAL EMPOWERMENT)

   Positively. Education, or improvement of local organizations (SOCIAL EMPOWERMENT)

   Positively. Financial resources, increased control of subsistence resources (ECONOMIC EMPOWERMENT)

   Positively, in OTHER WAYS (explain under Question 9)

   Negatively (explain under Question 9)

   Do not know

   Other _____

7. **What are the sources of financial support?**

   Government agency
   Civil society organization/NGO/community-based organization
   Private foundation
   Academic institution
   Do not know
   Other _____

8. **Which stages of the monitoring process were the community members and external agents (scientists, government staff) involved in?**

   Community members: The DESIGN of the monitoring system

   Community members: The DATA COLLECTION in the monitoring system

   Community members: The DATA INTERPRETATION in the monitoring system

   Community members: The USE OF THE RESULTS from the monitoring system

   External agents: The DESIGN of the monitoring system

   External agents: The DATA COLLECTION in the monitoring system

   External agents: The DATA INTERPRETATION in the monitoring system

   Do not know

   Other _____

9. **Anything else you want to add to the questions in this section?**

II. GENERAL INFORMATION

10. **Which attributes are monitored by the monitoring program?**

    Select all you agree with.

    WILDLIFE (animals)
    - Insects
    - Shellfish
    - Fish
    - Birds
    - Mammals
    - Other species

    VEGETATION
    - Fungi
    - Plants

    ABIOTIC PHENOMENA
    - Water
    - Air
    - Snow
    - Ice
    - Wind
    - Weather
    - Contaminants
    - Other abiotic phenomena

    SOCIOCULTURAL ATTRIBUTES
    - Human health
    - Wellness
    - Language
    - Traditional knowledge transmission
    - Other sociocultural attributes
    - Other

    _____

    _____

11. **What landscape type is monitored by the monitoring program?**

    Taiga or boreal forest
    Tundra
    Freshwater
    Coastal
    Sea
    Other _____

12. **Who decided on what and where data should be collected?**

    Scientists
    Government staff
    Community members
    Other _____

13. **Briefly describe the methodology (one sentence only, please).**

14. **Do you use some kind of measure of effort?**

    For example, number of hunting trips, hooks and net used?

    Yes
    No
    Do not know

15. **Equipment needs for data collection: How much does the equipment required by one data collector cost?**

    $ 0–100
    $ 100–1000
    $ > 1000
    Do not know

16. **What is the frequency of data collection?**

    Indicate the intervals between successive bouts of data collection.

    Daily
    Weekly
    Monthly
    Quarterly (three-monthly)
    Yearly
    Do not know
    Other _____

17. **Is monitoring done during certain time of year?**

    No distinct period
    January
    February
    March
    April
    May
    June
    July
    August
    September
    October
    November
    December
    Do not know

18. **Anything else you want to add to the questions in this section?**

III. COMMUNITY MEMBERS

19. **How many community members participate in total in the monitoring process?**

    Select one of the available choices.

    0–10
    11–100
    101–500
    501–1000
    >1000
    Do not know

20. **What is the proportion of women?**

    Select one of the available choices.

    0%
    1–5%
    6–25%
    26–75%
    76–100%
    Do not know

21. **What age classes are included in the monitoring process?**

    Children (0–18)
    Youth (19–26)
    Adults (27–60)
    Elders (>60)
    Do not know

22. **How were the community members chosen?**

    Appointed by somebody based on their background
    They proposed their involvement
    Do not know
    Other _____

23. **What are the sources of motivation for community members to participate in the monitoring system?**

    Have their voices heard/protect rights over land, sea and resources
    Leisure interest/socializing
    Social engagement
    Personal benefits
    Mandatory
    Sustain health and abundance of wildlife
    Do not know
    Other _____

24. **Do the community members get compensation/salary for being involved in the monitoring program?**

    Yes
    No
    Do not know

25. **Do community members obtain feedback on the findings from the monitoring?**

    Select one of the available choices.

    Yes
    No
    Do not know

26. **Anything else you want to add to the questions in this section?**

IV. THE DATA

27. **Are there data validation processes built into the monitoring program?**

    For example by triangulation (cross-checking)

    No inbuilt validation processes
    Yes, triangulation across COMMUNITY MEMBERS
    Yes, triangulation across COMMUNITIES
    Yes, triangulation across METHODS
    Yes, other types of validation processes (not triangulation)
    Do not know
    Other _____

28. Is the data quality of the collected data being checked?

For example, data spreadsheets can be checked for data encoding errors, anomalies, and data that are beyond expected range, before the data is used. Select one of the available choices.

Yes
No
Do not know
Other _____

29. What language is the original data in?

Select one of the available choices.

Local dialect (not the national language)
The national language
English
Do not know
Other _____

30. How long after data collection is the data available to users?

Select one of the available choices.

Data are accessible after an unknown period
Data are accessible some years after acquisition
Data are accessible within 6–12 months after acquisition
Data are accessible within a month after acquisition
Data are accessible within a week after acquisition
Data are accessible within a day after acquisition
Data are accessible within 3 hours after acquisition
Data are accessible in real time
Do not know
Other _____

31. Has any assessment of the program been undertaken within the last 3 years?

Select one of the available choices.

Yes
No
Do not know

32. Principal challenges to the monitoring program?

Select all you agree with.

Limited funding
Personal hardship
Violation of intellectual property rights/free prior and informed consent
Political challenges
Fatigue among community members
Do not know
Other _____

33. Anything else you would like to add about the monitoring program?

34. Your name (the encoder of the metadata)

35. Your email address

References

Adeyeye, Yemi, Shannon Hagerman, and Ricardo Pelai. 2019. "Seeking procedural equity in global environmental governance: Indigenous participation and knowledge politics in forest and landscape restoration debates at the 2016 World Conservation Congress." *Forest Policy and Economics* 109: 102006.

Alaska Arctic Policy Commission. 2015. *Final Report of the Alaska Arctic Policy Commission.* Juneau: Alaska Arctic Policy Commission. www.uaf.edu/caps/resources/state-of-alaska/aapc-final-report-2015.pdf

Albert, Thomas F. 2000. "The influence of Harry Brower, Sr., an Iñupiaq Eskimo hunter, on the bowhead whale research program conducted at the UIC-NARL facility by the North Slope Borough." In *Fifty More Years Below Zero*, edited by David Norton, 265–278. Arctic Institute of North America.

Arctic Council. 2009. *Arctic Marine Shipping Assessment 2009 Report.* Accessed June 19, 2020. www.pame.is/index.php/projects/arctic-marine-shipping/amsa

Armitage, Derek, Fikret Berkes, Aaron Dale, Erik Kocho-Schellenberg, and Eva Patton. 2011. "Co-management and the co-production of knowledge: Learning to adapt in Canada's Arctic." *Global Environmental Change* 21, no. 3: 995–1004.

Ashthorn, Heather. 2018. "Arctic Borderlands Lessons Learned." INTAROS Community-Based Monitoring Library. Accessed June 19, 2020. mkp28.wixsite.com/cbm-best-practice/arctic-borderlands-lessons-learned

Berkes, Fikret, and Derek Armitage. 2010. "Co-management institutions, knowledge, and learning: Adapting to change in the Arctic." *Études/Inuit/Studies* 34, no. 1: 109–131. doi.org/10.7202/045407ar

Bonney, Rick, Jennifer L. Shirk, Tina B. Phillips, Andrea Wiggins, Heidi L. Ballard, Abraham J. Miller-Rushing, and Julia K. Parrish. 2014. "Next steps for citizen science." *Science* 343, no. 6178: 1436–1437.

Brammer, Jeremy R., Nicolas D. Brunet, A. Cole Burton, Alain Cuerrier, Finn Danielsen, Kanwaljeet Dewan, Thora Martina Herrmann et al. 2016. "The role of digital data entry in participatory environmental monitoring." *Conservation Biology* 30, no. 6: 1277–1287.

Cochran, Patricia, Orville H. Huntington, Caleb Pungowiyi, Stanley Tom, F. Stuart Chapin, Henry P. Huntington, Nancy G. Maynard, and Sarah F. Trainor. 2013. "Indigenous frameworks for observing and responding to climate change in Alaska." In *Climate Change and Indigenous Peoples in the United States*, edited by Julie Koppel Maldonado, Benedict Colombi, and Rajul Pandya, 49–59. Switzerland: Springer, Cham.

Connors, John Patrick, Shufei Lei, and Maggi Kelly. 2012. "Citizen science in the age of neogeography: Utilizing volunteered geographic information for environmental monitoring." *Annals of the Association of American Geographers* 102, no. 6: 1267–1289.

Conrad, Cathy C., and Krista G. Hilchey. 2011. "A review of citizen science and community-based environmental monitoring: issues and opportunities." *Environmental Monitoring and Assessment* 176, no. 1–4: 273–291.

Convention on Biological Diversity. 2011. *Tkarihwai´e:ri code of ethical conduct to ensure respect for the cultural and intellectual heritage of indigenous and local communities relevant to the conservation and sustainable use of biological diversity*. Montreal: Secretariat of the Convention on Biological Diversity. Accessed June 19, 2020. www.cbd.int/traditional/code/ethicalconduct-brochure-en.pdf

Convention on Biological Diversity. 2019. Accessed June 19, 2020. www.cbd.int

Cuyler, Christine, Colin J. Daniel, Martin Enghoff, Nuka Møller-Lund, Nette Levermann, Per N. Hansen, Ditlev Damhus, and Finn Danielsen. 2020. "Using local ecological knowledge as evidence to guide management: A community-led harvest calculator for muskoxen in Greenland." *Conservation Science and Practice*. doi:10.1002/csp2.159

Danielsen, Finn, Neil D. Burgess, and Andrew Balmford. 2005. "Monitoring matters: Examining the potential of locally-based approaches." *Biodiversity & Conservation* 14, no. 11: 2507–2542.

Danielsen, Finn, Neil D. Burgess, Per M. Jensen, and Karin Pirhofer-Walzl. 2010. "Environmental monitoring: The scale and speed of implementation varies according to the degree of peoples involvement." *Journal of Applied Ecology* 47, no. 6: 1166–1168.

Danielsen, Finn, Per M. Jensen, Neil D. Burgess, Ronald Altamirano, Philip A. Alviola, Herizo Andrianandrasana, Justin S. Brashares et al. 2014a. "A multicountry assessment of tropical resource monitoring by local communities." *BioScience* 64, no. 3: 236–251.

Danielsen, Finn, Per M. Jensen, Neil D. Burgess, Indiana Coronado, Sune Holt, Michael K. Poulsen, Ricardo M. Rueda et al. 2014b. "Testing focus groups as a tool for connecting indigenous and local knowledge on abundance of natural resources with science-based land management systems." *Conservation Letters* 7, no. 4: 380–389.

Danielsen, Finn, Elmer Topp-Jørgensen, Nette Levermann, Piitaaraq Løvstrøm, Martin Schiøtz, Martin Enghoff, and Paviarak Jakobsen. 2014c. "Counting what counts: Using local knowledge to improve Arctic resource management." *Polar Geography* 37, no. 1: 69–91.

Danielsen, Finn, Karin Pirhofer-Walzl, Teis P. Adrian, Daniel R. Kapijimpanga, Neil D. Burgess, Per M. Jensen, Rick Bonney et al. 2014d. "Linking public participation in scientific research to the indicators and needs of international environmental agreements." *Conservation Letters* 7, no. 1: 12–24.

Danielsen, Finn, Martin Enghoff, Eyðfinn Magnussen, Tero Mustonen, Anna Degteva, Kia K. Hansen, Nette Levermann, Svein D. Mathiesen, and Øystein Slettemark. 2017. "Citizen science tools for engaging local stakeholders and promoting local and traditional knowledge in landscape stewardship." In *The Science and Practice of Landscape Stewardship*, edited by Claudia Bieling and Tobias Plieninger, 80–98. Cambridge: Cambridge University Press.

Danielsen, F., et al. n.d. "The concept, practice, application and results of locally-based monitoring of the environment." *BioScience*. In review.

Deemer, Gregory J., Uma S. Bhatt, Hajo Eicken, Pamela G. Posey, Jennifer K. Hutchings, James Nelson, Rebecca Heim, Richard A. Allard, Helen Wiggins, and Kristina Creek. 2017. "Broadening the sea-ice forecaster toolbox with community observations: A case study from the northern Bering Sea." *Arctic Science* 4, no. 1: 42–70. doi:10.1139/as-2016-0054

Díaz, Sandra, Josef Settele, Eduardo S. Brondízio, Hien T. Ngo, John Agard, Almut Arneth, Patricia Balvanera et al. 2019. "Pervasive human-driven decline of life on Earth points to the need for transformative change." *Science* 366, no. 6471. doi:10.1126/science.aax3100

Dowsley, Martha, Shari Gearheard, Noor Johnson, and Jocelyn Inksetter. 2010. "Should we turn the tent? Inuit women and climate change." *Études/Inuit/Studies* 34, no. 1: 151–165.

Eerkes-Medrano, Laura, David E. Atkinson, Hajo Eicken, Bill Nayokpuk, Harvey Sookiayak, Eddie Ungott, and Winton Weyapuk Jr. 2017. "Slush-ice berm formation on the west coast of Alaska." *Arctic* 70, no. 2: 190–202.

Eicken, Hajo. 2010. "Indigenous knowledge and sea ice science: What can we learn from indigenous ice users?" In *SIKU: Knowing Our Ice—Documenting Inuit Sea Ice Knowledge and Use,* edited by Igor Krupnik, Claudio Aporta, Shari Gearheard, Gita J. Laidler, and Lene Kielsen Holm, 357–376. New York: Springer-Verlag.

Eicken, Hajo, Finn Danielsen, Josie Sam, Maryann Fidel, Noor Johnson, Michael K. Poulsen, Olivia A. Lee et al. n.d. "Connecting top-down and bottom-up approaches in environmental observing." *BioScience*. In review.

Eicken, Hajo, Mette Kaufman, Igor Krupnik, Peter Pulsifer, Leonard Apangalook, Paul Apangalook, Winton Weyapuk Jr, and Joe Leavitt. 2014. "A framework and database for community sea ice observations in a changing Arctic: An Alaskan prototype for multiple users." *Polar Geography* 37, no. 1: 5–27.

Eicken, Hajo, Liesel A. Ritchie, and Ashly Barlau. 2011. "The role of local, indigenous knowledge in Arctic offshore oil and gas development, environmental hazard mitigation, and emergency response." In *North by 2020*, edited by Amy Lovecraft and Hajo Eicken, 577–603. Fairbanks: University of Alaska Press.

EMAN (The Ecological Monitoring and Assessment Network). 2003. *Improving Local Decision Making through Community Based Monitoring: Toward a Canadian Community Monitoring Network.* Ottawa: Environment Canada. publications.gc.ca/site/eng/9.699150/publication.html

Enghoff, Martin, Nikita Vronski, Vyacheslav Shadrin, Rodion Sulyandziga, and Finn Danielsen. 2019. *INTAROS Community-Based Monitoring Capacity Development Process in Yakutia and Komi Republic, Arctic Russia.* CSIPN, RIPOSR, NORDECO and INTAROS.

Etiendem, Denis N., Rebecca Jeppesen, Jordan Hoffman, Kyle Ritchie, Beth Keats, Peter Evans, and Danielle E. Quinn. 2020. "The Nunavut Wildlife Management Board's community-based monitoring network: Documenting Inuit harvesting experience using modern technology." *Arctic Science* 6, no. 3: 307–325.

Eynden, Veerle Van den, Louise Corti, Matthew Woollard, Libby Bishop, and Laurence Horton. 2011. *Managing and Sharing Data. Best Practice for Researchers.* Essex, UK: University of Essex Press. ukdataservice.ac.uk/media/622417/managingsharing.pdf

Fidel, Maryann, Noor Johnson, Finn Danielsen, Hajo Eicken, Lisbeth Iversen, Olivia Lee, and Colleen Strawhacker. 2017. *INTAROS Community-based Monitoring Experience Exchange Workshop Report.* Fairbanks: Yukon River Inter-Tribal Watershed Council (YRITWC), University of Alaska Fairbanks, ELOKA, and INTAROS.

Ford, J. D., L. Berrang-Ford, R. Biesbroek, M. Araos, S. E. Austin, and A. Lesnikowski. 2015. "Adaptation tracking for a post-2015 climate agreement." *Nature Climate Change* 5, no. 11: 967–969.

Gearheard, Shari, and Jamal Shirley. 2007. "Challenges in Community-Research Relationships: Learning from Natural Science in Nunavut." *Arctic* 60, no 1: 62–74.

Gofman, Victoria. 2010. *Community Based Monitoring Handbook: Lessons from the Arctic. CAFF CBMP Report No. 21.* Akureyri, Iceland: Conservation of Arctic Flora and Fauna. Accessed June 19, 2020. oaarchive.arctic-council.org/bitstream/handle/11374/210/CBM_Handbook_Oct_2011.pdf

Green, Jessica F. 2005. "Engaging civil society in global governance: Lessons for member states and advice for civil society." *Global Environmental Change* 15, no. 1: 69–72.

Griffith, David L., Lilian Alessa, and Andrew Kliskey. 2018. "Community-based observing for social–ecological science: lessons from the Arctic." *Frontiers in Ecology and the Environment* 16, no. S1: S44–S51. doi.org/10.1002/fee.1798

Herman-Mercer, Nicole, Ronald Antweiler, Nicole Wilson, Edda Mutter, Ryan Toohey, and Paul Schuster. 2018. "Data quality from a community-based, water-quality monitoring project in the Yukon River Basin." *Citizen Science: Theory and Practice* 3, no. 2. doi.org/10.5334/cstp.123

Hovelsrud, Grete K., and Charlotte Winsnes (eds.). 2006. Conference on User Knowledge and Scientific Knowledge in Management Decision-Making, Reykjavík, Iceland, 4-7 January 2003. Conference Proceedings. NAMMCO. 95pp. nammco.no/wp-content/uploads/2016/10/total-publication.pdf

Huntington, Henry P. 1998. "Observations on the utility of the semi-directive interview for documenting traditional ecological knowledge." *Arctic* 51: 237–242.

Huntington, Henry P. 2011. "The local perspective." *Nature* 478: 182–183.

Huntington, Henry, Finn Danielsen, Martin Enghoff, Nette Levermann, Peter Løvstrøm, Martin Schiøtz, Michael Svoboda, and Elmer Topp-Jørgensen. 2013. "Conservation through community involvement." In *Arctic Biodiversity Assessment: Status and Trends in Arctic Biodiversity*, edited by Hans Meltofte, 644–647. Akureyi, Iceland: Conservation of Arctic Flora and Fauna. www.caff.is/assessment-series/233-arctic-biodiversity-assessment-2013/download

IASC. 2013. "Statement of Principles and Practices for Arctic Data Management." April 16, 2013. International Arctic Science Committee. Accessed June 19, 2020. iasc.info/data-observations/iasc-data-statement

INTAROS. 2020. "Concept and objectives." Accessed June 19, 2020. www.intaros.eu/about/concept-objectives/

Interagency Arctic Policy Research Committee. 2018. "Principles for conducting research in the Arctic." US Interagency Arctic Research Policy Committee. Accessed June 19, 2020. www.iarpccollaborations.org/uploads/cms/documents/principles_for_conducting_research_in_the_arctic_final_2018.pdf

Inuit Circumpolar Council. n.d. "Application of Indigenous Knowledge within the Arctic Council." Accessed June 19, 2020. iccalaska.org/wp-icc/wp-content/uploads/2016/03/Application-of-IK-in-the-Arctic-Council.pdf

Inuit Circumpolar Council. 2018. *Wildlife Management Summit Report*. Ottawa, Canada. secureservercdn.net/104.238.71.250/hh3.0e7.myftpupload.com/wp-content/uploads/ICC-Wildlife-Management-Summit-Report.pdf

Inuit Circumpolar Council—Alaska. 2015. *Alaskan Inuit Food Security Conceptual Framework: How to Assess the Arctic from an Inuit Perspective*. Anchorage: Inuit Circumpolar Council—Alaska.

Inuit Circumpolar Council—Alaska. 2020. *Food Sovereignty and Self-Governance: Inuit Role in Managing Arctic Marine Resources*. Anchorage: Inuit Circumpolar Council—Alaska. Accessed November 10, 2020. iccalaska.org/wp-icc/wp-content/uploads/2020/09/FSSG-Report_-LR.pdf

IPBES. 2019. "Summary for policymakers of the global assessment report on biodiversity and ecosystem services of the Intergovernmental Science-Policy Platform on Biodiversity and Ecosystem Services." Edited by Sandra Díaz, Josef Settele, Eduardo Brondízio, Hien T. Ngo, Maximilien Guèze, John Agard, Almut Arneth et al. Bonn, Germany: IPBES Secretariat.

IPCC. 2019. "Summary for policymakers." In *IPCC Special Report on the Ocean and Cryosphere in a Changing Climate*, edited by Hans-Otto Pörtner, Debra C. Roberts, Valérie Masson-Delmotte, Panmao Zhai, M. Tignor, Elvira Poloczanska, Katja Mintenbeck et al. In press. www.ipcc.ch/site/assets/uploads/sites/3/2019/11/03_SROCC_SPM_FINAL.pdf

Johnson, Mark, Hajo Eicken, Matthew L. Druckenmiller, and Richard Glenn, eds. 2014. *Experts workshops to comparatively evaluate coastal currents and ice movement in the northeastern Chukchi Sea; Barrow and Wainwright, Alaska. March 11–15, 2013.* Fairbanks: University of Alaska Fairbanks. www.uaf.edu/cfos/files/research/projects/experts-workshops-to-comp/Report-Final---29APR14.pdf

Johnson, Noor, Lilian Alessa, Carolina Behe, Finn Danielsen, Shari Gearheard, Victoria Gofman-Wallingford, Andrew Kliskey et al. 2015. "The contributions of community-based monitoring and traditional knowledge to Arctic observing networks: Reflections on the state of the field." *Arctic* 68, Suppl. 1: 28–40.

Johnson, Noor, Carolina Behe, Finn Danielsen, Eva M. Krümmel, Scot Nickels, and Peter L. Pulsifer. 2016. *Community-based Monitoring and Indigenous Knowledge in a Changing Arctic: A Review for the Sustaining Arctic Observing Networks.* Sustaining Arctic Observing Networks. Ottawa: Inuit Circumpolar Council.

Johnson, Noor, Maryann Fidel, Finn Danielsen, Lisbeth Iversen, Michael K. Poulsen, Donna Hauser, and Peter Pulsifer. 2018. *INTAROS Community-based Monitoring Experience Exchange Workshop Report Québec City, Québec.* ELOKA, Yukon River Inter-Tribal Watershed Council (YRITWC), University of Alaska Fairbanks, and INTAROS, Québec City, Québec.

Johnson, Noor, Matthew L. Druckenmiller, Finn Danielsen, and Peter L. Pulsifer. "The Use of Digital Platforms for Community-Based Monitoring". BioScience". In review.

Jones, Julia P. G, Ben Collen, Giles Atkinson, Peter WJ Baxter, Philip Bubb, Janine B. Illian, Todd E. Katzner et al. 2011. "The why, what, and how of global biodiversity indicators beyond the 2010 target." *Conservation Biology* 25: 450–457.

Kanie, Norichika. 2007. "Governance with multilateral environmental agreements: A healthy or ill-equipped fragmentation?" In *Global Environmental Governance,* edited by Walter Hoffmann and Lydia Swart, 67–86. New York: Center for UN Reform Education.

Kendall, James J., Jeffrey J. Brooks, Chris Campbell, Kathleen Wedemeyer, Catherine Coon, Sharon Warren, Guillermo Auad et al. 2017. "Use of traditional knowledge by the United States Bureau of Ocean Energy Management to support resource management." *Czech Polar Rep* 7: 151–163.

Kouril, Diana, Chris Furgal, and Tom Whillans. 2015. "Trends and key elements in community-based monitoring: a systematic review of the literature with an emphasis on Arctic and Subarctic regions." *Environmental Reviews* 24, no. 2: 151–163. doi.org/10.1139/er-2015-0041

Kutz, Susan, and Matilde Tomaselli. 2019. "'Two-eyed seeing' supports wildlife health." *Science* 364, no. 6446: 1135–1137. doi.org/10.1126/science.aau6170

Latham, Robert, and Lisa Williams. 2013. "Power and inclusion: Relations of knowledge and environmental monitoring in the Arctic." *Journal of Northern Studies* 7, no. 1: 7–30.

Lovecraft, Amy L., et al. n.d. "Ecological democracy and planning for resilience: Evaluating social learning in regional scenarios development for Arctic Alaska." *Journal of Environmental Policy and Planning.* In review.

Magnussen, Eyðfinn. 2016. "Hunting of hare in the Faroe Islands in 2015" (in Faroese). *Haruveiðan í Føroyum 2015*. NVDrit 2016:02: 1–44. Tórshavn: University of the Faroe Islands. Accessed June 19, 2020. www.setur.fo/media/5017/nvdrit2016_02_haruveidanifoeroyum2015.pdf

Mahoney, Andrew, and Shari Gearheard. 2008. "Handbook for community-based sea ice monitoring." NSIDC Special Report 14. Boulder, Colorado: National Snow and Ice Data Center.

McIntyre, L. 1989. "Last Days of Eden." *National Geographic* 174, no. 6: 800–817.

Mercer, Jessica, Ilan Kelman, Lorin Taranis, and Sandie Suchet-Pearson. 2010. "Framework for integrating indigenous and scientific knowledge for disaster risk reduction." *Disasters* 34, no. 1: 214–239. doi:10.1111/j.1467-7717.2009.01126.

Ministry of Fisheries, Hunting and Agriculture. 2011. "Hvilke muligheder har bygder og kommuner for at tage forvaltningsbeslutninger om levende ressourcer af relevans for fangere, fiskere og naturinteresserede i Akunnaaq, Qaarsut og Ilulissat?" pisuna.org/documents/Rapport%20om%20muligheder%20for%20lokal%20forvaltning%20DK.docx

Mistry, Jayalaxshmi, and Andrea Berardi. 2016. "Bridging indigenous and scientific knowledge." *Science* 352, no. 6291: 1274–1275.

Mitchell, Ronald B. 2003. "International environmental agreements: A survey of their features, formation, and effects." *Annual Review of Environment and Resources* 28, no. 1: 429–461.

Moller, Henrik, Fikret Berkes, Philip O'Brian Lyver, and Mina Kislalioglu. 2004. "Combining science and traditional ecological knowledge: Monitoring populations for co-management." *Ecology and Society* 9, no. 3. www.ecologyandsociety.org/vol9/iss3/art2/

Mustonen, Tero. 2014. "Power discourses of fish death: Case of Linnunsuo peat production." *Ambio* 43, no. 2: 234–243.

Mustonen, Tero, Vladimir Feodoroff, Pauliina Feodoroff, Aqqalu Olsen, Per Ole Fredriksen, Kaisu Mustonen, Finn Danielsen, Nette Levermann, Augusta Jeremiassen, Helle T. Christensen et al. 2018. *Deepening Voices—eXchanging Knowledge of Monitoring Practices between Finland and Greenland*. SnowChange Cooperative, NORDECO, KNAPK, Greenland Institute of Natural Resources, and Greenland Ministry of Fisheries and Hunting. Accessed June 19, 2020. www.snowchange.org/pages/wp-content/uploads/2018/01/gronlanti.pdf

Mutter, Edda, and Maryann Fidel. 2018. "What worked, and what didn't: Lessons learned from the Indigenous Observation Network." INTAROS Community-Based Monitoring Library. Accessed June 19, 2020. mkp28.wixsite.com/cbm-best-practice/ion-yukon-river-lessons-learned

Nadasdy, Paul. 1999. "The politics of TEK: Power and the 'integration' of knowledge." *Arctic Anthropology* 36, no. 1–2: 1–18.

Nickels, Scot, and Cathleen Knotsch. 2011. "Inuit perspectives on research ethics: The work of Inuit Nipingit." *Études/Inuit/Studies* 35, no. 1–2: 57–81.

Nordic Council of Ministers. 2015. *Local Knowledge and Resource Management: On the Use of Indigenous and Local Knowledge to Document and Manage Natural Resources in the Arctic.* TemaNord 2015:506. Copenhagen: Nordic Council of Ministers.

Parlee, Brenda, and Lutsel K'e Dene First Nation. 1998. *A Guide to Community-Based Monitoring for Northern Communities.* Northern Minerals Program Working Paper No. 5. Canadian Arctic Resources Committee. Accessed June 19, 2020. www.carc.org/wp-content/uploads/2017/10/NMPWorkingPaper5Parlee.pdf

Parsons, Mark A., Øystein Godøy, Ellsworth LeDrew, Taco F. De Bruin, Bruno Danis, Scott Tomlinson, and David Carlson. 2011. "A conceptual framework for managing very diverse data for complex, interdisciplinary science." *Journal of Information Science* 37, no. 6: 555–569.

Peachey, Karen. 2015. "On-the-ground Indigenous stewardship programs across Canada: Inventory project." Accessed June 19, 2020. www.indigenousguardianstoolkit.ca/sites/default/files/Community%20Resource_Final%20Report%20with%20Profiles%20March%2027%202015_1.pdf

Pecl, Gretta T., Miguel B. Araújo, Johann D. Bell, Julia Blanchard, Timothy C. Bonebrake, I-Ching Chen, Timothy D. Clark et al. 2017. "Biodiversity redistribution under climate change: Impacts on ecosystems and human well-being." *Science* 355, no. 6332: eaai9214.

Posey, Darrell Addison. 1998. "Biodiversity, genetic resources and indigenous peoples in Amazonia: (re) discovering the wealth of traditional resources of native Amazonians." *Amazonia 2000: Development, Environment and Geopolitics.* 24–26 June 1998. London: Institute of Latin American Studies, University of London.

Poulsen, Michael K., Lisbeth Iversen, Naja E. Mikkelsen, and Finn Danielsen. 2019. *Cruise Expedition Monitoring Workshop and Dialogue-Seminar: On Improving and Expanding the Environmental Monitoring Efforts of Cruise Ships in the Arctic.* Bergen: NORDECO, NERSC and INTAROS.

Protection of the Arctic Marine Environment (PAME). 2017. *Meaningful Engagement of Indigenous Peoples and Local Communities in Marine Activities Report Part I: Arctic Council and Indigenous Engagement—A Review.* Arctic Council. Akureyri: Protection of the Arctic Marine Environment Secretariat.

Pulsifer, Peter. 2015. "Indigenous knowledge: Key considerations for Arctic research and data management." *Sharing Knowledge: Traditions, Technologies, and Taking Control of Our Future Workshop. Exchange for Local Observations and Knowledge of the Arctic (ELOKA).* 22–24 September 2015. Boulder, Colorado. Accessed June 19, 2020. www.arcticobservingsummit.org/sites/arcticobservingsummit.org/files/Pulsifer-ELOKA--Extended_Sharing_Knowledge_statement.pdf

Pulsifer, Peter L., Gita J. Laidler, Fraser Taylor, and Amos Hayes. 2011. "Towards an Indigenist data management program: Reflections on experiences developing an atlas of sea ice knowledge and use." *The Canadian Geographer/Le Géographe canadien* 55, no. 1: 108–124.

Pulsifer, Peter, Shari Gearheard, Henry P. Huntington, Mark A. Parsons, Christopher McNeave, and Heidi S. McCann. 2012. "The role of data management in engaging communities in Arctic research: Overview of the Exchange for Local Observations and Knowledge of the Arctic (ELOKA)." *Polar Geography* 35, no. 3–4: 271–290.

Raynolds, Laura T., and Elizabeth A. Bennett. eds. 2015. *Handbook of Research on Fair Trade*. Cheltenham, UK: Edward Elgar Publishing.

SAON. n.d. "Sustaining Arctic Observing Networks Implementation." Accessed June 19, 2020. www.arcticobserving.org/images/pdf/Strategy_and_Implementation/SAON_Implementation_Plan_version_17JUL2018_Status_approved.pdf

Scassa, Teresa, Nate J. Engler, and DR Fraser Taylor. 2015. "Legal issues in mapping traditional knowledge: Digital cartography in the Canadian north." *The Cartographic Journal* 52, no. 1 (2015): 41–50.

Sigman, Marilyn. 2015. *Community-Based Monitoring of Alaska's Coastal and Ocean Environment: Best Practices for Linking Alaska Citizens with Science*. Fairbanks: Alaska Sea Grant, University of Alaska Fairbanks. doi.org/10.4027/cbmacoe.2015

Slats, Richard, Carol Oliver, Robert Bahnke, Helen Bell, Andrew Miller, Delbert Pungowiyi, Jacob Merculief et al. 2019. "Voices from the front lines of the changing Bering Sea: An Indigenous perspective to the 2019 Arctic Report Card." *Arctic Report Card*. National Oceanic and Atmospheric Administration. Accessed June 19, 2020. arctic.noaa.gov/Report-Card/Report-Card-2019/ArtMID/7916/ArticleID/850/Voices-from-the-Front-Lines-of-a-Changing-Bering-Sea

Tengö, Maria, Rosemary Hill, Pernilla Malmer, Christopher M. Raymond, Marja Spierenburg, Finn Danielsen, Thomas Elmqvist, and Carl Folke. 2017. "Weaving knowledge systems in IPBES, CBD and beyond—lessons learned for sustainability." *Current Opinion in Environmental Sustainability* 26: 17–25.

Tengö, Maria, B. Austin, Finn Danielsen, and Á. Fernández-Llamazares. n.d. "Citizen science and Indigenous and Local Knowledge Systems." *BioScience*. In review.

UArctic. 2020. "Thematic Network on Collaborative Resource Management." Accessed June 19, 2020. www.uarctic.org/organization/thematic-networks/collaborative-resource-management/

United Nations. 2008. "United Nations Declaration on the Rights of Indigenous Peoples." Accessed June 19, 2020. www.un.org/development/desa/indigenouspeoples/declaration-on-the-rights-of-indigenous-peoples.html

Wagner, Penelope, Nich Hughes, Edda Falk, Lisa Kelley, Alex Cowan et al. n.d. "Evolving polar tourism vessel requirements for environmental data and the role of citizen science." In review.

Ward-Fear, Georgia, Balanggarra Rangers, David Pearson, Melissa Bruton, and Rick Shine. 2019. "Sharper eyes see shyer lizards: Collaboration with indigenous peoples can alter the outcomes of conservation research." *Conservation Letters*: e12643.

Wheeler, Helen C., Dominique Berteaux, Chris Furgal, Brenda Parlee, Nigel G. Yoccoz, and David Grémillet. 2016. "Stakeholder perspectives on triage in wildlife monitoring in a rapidly changing Arctic." *Frontiers in Ecology and Evolution* 4: 128. doi.org/10.3389/fevo.2016.00128

Wheeler, Helen, Finn Danielsen, Maryann Fidel, Vera Hausner, Tim Horstkotte, Noor Johnson, Olivia Lee et al. 2020. "The need for transformative changes in the use of Indigenous knowledge along with science for environmental decision-making in the Arctic". *People and Nature* 2: 544–546.

White House. 2016a. "Joint Statement of Ministers on the Occasion of the first White House Arctic Science Ministerial." Accessed June 19, 2020. obamawhitehouse.archives.gov/the-press-office/2016/09/28/joint-statement-ministers

White House. 2016b. "United States-Canada Joint Arctic Leaders' Statement." Accessed June 19, 2020. obamawhitehouse.archives.gov/the-press-office/2016/12/20/united-states-canada-joint-arctic-leaders-statement

Wilkinson, Mark D., Michel Dumontier, Ijsbrand Jan Aalbersberg, Gabrielle Appleton, Myles Axton, Arie Baak, Niklas Blomberg et al. 2016. "The FAIR Guiding Principles for Scientific Data Management and Stewardship." *Scientific Data* 3 (March): 160018. www.nature.com/articles/sdata201618

Wilson, Nicole. 2017. *Indigenous Observation Network: Evaluating Community-Based Water Quality Monitoring in the Yukon River Basin*. Vancouver: University of British Columbia, Institute for Resources, Environment and Sustainability.

Young-Ing, Greg. 2008. "Conflicts, discourse, negotiations and proposed solutions regarding transformation of traditional knowledge." In *Aboriginal Oral Traditions: Theory, Practice, Ethics,* edited by Renate Eigenbrod and Renee Hulan, 61–78. Toronto: Brunswick Books.

Index

Note: page numbers followed by *f*, *t* to *b* refer to figures, tables and boxes respectively.

A

ABES. *See* Arctic Borderlands Ecological Knowledge Society
Agreement on the Conservation of African-Eurasian Migratory Waterbirds (AEWA), 23*t*
Agreement to Prevent Unregulated High Seas Fisheries in the Central Arctic Ocean, 28
Alaska Arctic Observatory and Knowledge Hub (A-OK), 23*t*, 42, 54, 87*t*
Älgdata, 23*t*, 89*t*
A-OK. *See* Alaska Arctic Observatory and Knowledge Hub
Arctic, challenging features for CBM, 27
Arctic and Earth SIGNs (Alaska), 23*t*, 54–56, 87*t*
Arctic Borderlands Ecological Knowledge Society (ABES) [Canada], 23*t*, 39, 53–54, 87*t*
Arctic Data Explorer, 42, 44*t*
Arctic Eider Society (Canada), 34, 53
Arctic Report Card (2019), contribution of Indigenous knowledge to, 82–83
Arctic Science Ministerial meeting (2016), 28
assessment of Arctic CBM programs. *See also* questionnaire; workshops with CBM participants
    areas assessed, 4–5
    contributions of this study, 84
    detailed analysis, need for, 9
    methods of, 4–7
    number identified, ix, 4, 9–10
    number selected for study, ix, 4, 9
    as part of INTAROS study, 9
    previous reviews, developments since, 28
    process for, 5
    sampling error and, 9–10
Atlas of Community-Based Monitoring and Indigenous Knowledge in a Changing Arctic, 14
atmosphere and weather, groups monitoring, 14

B

Bern Convention on the Conservation of European Wildlife and Natural Habitats, 23*t*
Bird Phenology, 23*t*, 87*t*
BuSK, 23*t*, 87*t*

C

Calista Education and Culture, 34–35
CARE principles for data sharing, 45
CBD. *See* Convention on Biological Diversity

CBM. *See* community-based monitoring
CBMP. *See* Inuvialuit Settlement Region Community-Based Monitoring Program
Centre for Support to Indigenous Peoples of the North (Russia), 30*b*, 53
challenges for Arctic CBM programs, xi, 21
    data format incompatibility, 68–70, 72, 76, 83
    failure to represent community priorities, 64–66, 75
    funding as, 21, 26, 73, 84
    Indigenous rights protections, 70–73, 76, 84
    internet access, 83
    management agencies' lack of interest in CBM input, xi, 60–63, 75
    misconceptions about data quality, xi, 46, 61, 82
    organizational and support structures, 73–75, 76
    participant engagement, 66–68, 75
    sustainability, 21, 84
CITES. *See* Convention on International Trade in Endangered Species of Wild Fauna and Flora
Citizen Science Association Conference (North Carolina, 2019), 83–84
climate change
    impact on Arctic communities, 1, 40*f*
    and value of CBM data, 1–2, 83
CMS. *See* Conservation of Migratory Species agreement
communication
    between all program partners, importance of, 75
    between management and participants, importance of, 32
communities. *See also* Indigenous people of Arctic
    collaboration with, in program design, 28–31, 30*b*, 56, 61, 65, 71
    impact of climate change on, 40*f*
    incorporating data collection into everyday activities of, 36, 38, 67
    management agencies' lack of interest in input from, 60–63, 75
    and outside agents, balancing interests of, 84–85
    positive impact of CBM programs on, x, 12–13, 24, 30*b*, 42, 53, 67*f*
    priorities of, CBM programs not representing, 64–66, 75
    protecting rights of, 53–57, 58
    tailoring of CBM to, 31–32, 57
    varying interests and skills within, 31

community-based monitoring (CBM)
    definitions of, 7, 8*b*
    growing interest in, 76
    history of, 10, 10*f*, 79
    vs. scientist-executed monitoring
        asymmetric power issues and, 45
        as complementary, x, 25, 43
        and data format incompatibility, 68–70, 72, 76, 83
        differences in method, 8*b*
        equivalent validity of data, xi, 46, 49–51, 50*t*
Community-based Monitoring and Indigenous Knowledge in a Changing Arctic (Johnson et al.)
    as basis of this study, ix, 3
    on CBM good practices, 28
    current study's contribution and, 4
    definition of CBM, 7, 8*b*
    workshops used in, 4
community-based monitoring programs (CBM) in Arctic, 87*t*–89*t*. *See also* assessment of Arctic CBM programs; challenges for Arctic CBM programs; future of Arctic CBM; good practice
    age of, 10, 10*f*, 24
    attributes monitored by, 13, 14*f*
    benefits of, need for publicizing, 82–83
    biome types monitored by, x, 16, 17*f*, 25–26
    characteristics of, 10–13
    and community goals, positive impact on, x, 12–13, 24, 30*b*, 42, 53, 67*f*
    domains observed in, ix–x, 13–14, 25
    geographic distribution, x, 14–16, 15*f*, 25
    groups involved in, ix, 2
    importance of, 68, 77
    linkages to key economic sectors, 16–17, 18*t*, 26, 29
    locations of, 5*f*
    methods used by, diversity of, x, xi
    presentations to young people on, 78*f*
    special capabilities of, 43, 46, 51
    temporal coverage, 16, 17*f*
    types of programs included under, 7, 8*b*
Conservation of Migratory Species agreement (CMS), 23*t*
Convention for the Protection of the Marine Environment of the North-East Atlantic (OSPAR), 23*t*
Convention on Biological Diversity (CBD), 22, 23*t*
Convention on International Trade in Endangered Species of Wild Fauna and Flora (CITES), 23*t*, 79

Convention on Persistent Organic Pollutants (POP), 23*t*
costs of Arctic CBM programs, 21. *See also* funding for Arctic CBM
cruise ships, expeditionary
　environmental impact, need for monitoring of, 82
　as potential sources of data, 81–82
CSIPN (Center for Support to Indigenous Peoples of the North), 23*t*, 87*t*

D

data from Arctic CBM
　accessibility
　　good practice in, 42–43, 58
　　importance of, 19, 34
　　restrictions on access, 54, 58
　and advocacy-based evidence *vs.* evidence-based advocacy, 61
　collection of, 7*f*, 11*f*, 29*f*, 43*t*, 63*f*, 72*f*
　　effort needed for, management awareness of, 32
　　incorporation into participants' everyday activities, 36, 38, 67
　　methods of, 19, 25, 47–48
　　planning of, 29
　　training in, 30*b*, 32
　expeditionary cruise ships and, 81–82
　format incompatibility
　　causes and effects of, 68–69, 72, 83
　　collection method and, 19
　　interventions to improve, 69–70
　interpretation, training in, 30*b*, 32
　knowledge products
　　benefits to decision-makers, 41
　　coordinating with scientist-executed monitoring programs, 41–42, 45, 57–58
　　feeding back to participants, 34, 35, 35*f*
　　good practice in assuring quality of, 46–52, 48*t*, 58
　　incentives for government agencies' use of, 62
　　procedures to assure impact of, 38–41
　　providing to decision-makers, 36
　　time to availability of, 19
　misconceptions about poor quality of, xi, 46, 61, 82
　ownership and use, need for agreements on, 71
　ownership rights to, 53–54, 57, 58, 71
　public databases, 21, 26
　qualitative, percentage of programs collecting, 19
　quantitative, percentage of programs collecting, 19
　reporting of, x, xi
　sensitive, proper treatment of, 70
　storage and preservation, 20–21, 26, 33–34, 69 (*See also* web-based knowledge management platforms)
　　in global data repositories, 41, 42–45, 44*t*, 58
　　need for prior agreement on, 71
　uses, planning of, 29, 31
　validity of
　　and bias, 46–48, 52
　　potential source of inaccuracy, 46–48
　　procedures for assuring, 19, 46–52
　　and sampling protocols, collaboration in developing, 49
　　vs. scientist-executed monitoring, xi, 43, 46, 49–51, 50*t*
　value of, x, xi, 1–2, 43, 52*f*, 85
data sharing
　advances in, 28
　CARE principles and, 45
　FAIR principles and, 44–45
　global data repositories and, 41, 42–45, 44*t*, 58
　incompatibility of data format and, 68–70, 72, 76, 83
　reasons for limiting, 45
　value of, CBM participants' emphasis on, 44–45
data sovereignty, 45, 83
design of CBM programs. *See* establishing CBM programs
digital communication tools, and participant engagement, 33

E

eco labeling, use of CBM for, 80
economic sectors, key, linkage of Arctic CBM to, 16–17, 18*t*, 26, 29
economic value of CBM, making use of, 80
ELOKA. *See* Exchange for Local Observations and Knowledge of the Arctic
environmental goals. *See* international environmental agreements
Eskimo Whaling Commission (Alaska), 60–61
establishing CBM programs
　case study of, 30*b*
　collaboration with communities in, 28–31, 30*b*, 56, 61, 65, 71

establishing CBM programs (*continued*)
  good practice in, 28–32, 57
European Charter on Participatory Democracy in Spatial Planning Processes, 77
European Union
  Arctic policy of, 22–24
  Horizon 2020 Program, 2
  principle of subsidiarity, 48
Evenk & Izhma Peoples, 23*t*, 87*t*
everyday activities, 59*f*
  incorporating data collection into, 36, 38, 67
Exchange for Local Observations and Knowledge of the Arctic (ELOKA), 20–21, 34, 54, 58
exchanges between CBM programs, value of, 33

F

Facebook
  for communication between management and participants, 34, 41
  and feedback of knowledge products to participants, 35*f*
FAIR principles for data sharing, 44–45
Farmers and Herders. *See* Summer Farmers and Small Herders
Fávllis, 23*t*, 87*t*
Federation of Icelandic River Owners (Iceland), 10, 23*t*, 39, 42, 89*t*
Finnish Meteorological Institute (FMI) Snow Depth, 10, 23*t*, 88*t*
food, Indigenous dependence on natural environment for, 65*f*
FPIC. *See* free, prior and informed consent
free, prior and informed consent (FPIC), xi, 45, 53, 57, 58, 71, 76, 84
Fuglavernd, 23*t*, 88*t*
funding for Arctic CBM
  as challenge, 21, 26, 73, 84
  cost per data collector, 21
  long-term, programs with, 20
  as source for partnerships and support, 74
  sources of, 20, 25
funding for workshops with CBM participants, 2
future of Arctic CBM, 77–85
  and balancing of community and outsider interests, 84–85
  and climate change, value of CBM data on, 1–2, 83
  contributions of this study to, 84
  data compatibility and, 83
  and decentralized resource management, 79–80
  further research, areas requiring, 82–85
  growing environmental concerns and, 77
  improvement, areas requiring, 79–82
  increased Internet access, need for, 83
  innovative uses, need for, 80
  and local and Indigenous knowledge, increased engagement with, 79
  and local stakeholder inclusion, growing emphasis on, 77–78
  and managing staff, need for better training of, 80–81
  and monitoring of cruise ship impact, 82
  and participant engagement, research on, 83–84, 85
  remote locations, need for increased monitoring in, 81–82
  and socially-sustainable development, promotion of, 80
  and value of CBM, publicizing of, 78–79, 82–83
  and web-based knowledge management, potential for expansion of, 79

G

gender
  and differences in interests, 31
  of participants in Arctic CBM, 12, 24
Geomatics and Cartographic Research Centre (Carleton University), 34
George River, 23*t*, 88*t*
Global Indigenous Data Alliance, 45
Global Learning and Observations to Benefit the Environment program (GLOBE), 56
goals and values, shared, as key to smooth operation, 32
good practice
  in assuring knowledge product quality, 46–52, 48*t*, 58
  in coordinating with scientists' monitoring programs, 41–42, 45, 57–58
  in data accessibility, 42–43, 58
  definition of, 27
  in establishing CBM programs, 28–32, 30*b*, 57
  in implementing CBM programs, 32–36, 57, 61, 84
  for obtaining impact, 38–41, 57
  previous reviews, developments since, 28
  on protecting rights of Indigenous and local communities, 53–57, 58
  in sharing data with global data repositories, 41, 42–45, 44*t*, 58
  sources for information on, 27
  in sustaining CBM programs, 36–38, 57
Gordon Foundation, 34

H

halibut catch in Greenland, decentralized management of, 79–80
Hares, 23*t*, 88*t*

I

ice deserts, x, 16, 25–26
ICES. *See* International Council for the Exploration of the Sea
ICRW. *See* International Convention for the Regulation of Whaling
Imalirijiit program (Canada), 56
impact of CBM programs
　good practice for obtaining, 38–41, 57
　as incentive for participant engagement, 36, 38, 58, 66, 75, 84
　on resource management, x, xi, 19–20, 25, 26
implementing CBM programs, good practice in, 32–36, 57, 61, 84
Indigenous and local knowledge
　and Arctic Report Card (2019), 82–83
　increasing emphasis on, 28, 57, 77–78
　insufficient incorporation by management agencies, 79
　and intellectual property rights, 54, 57
　Inuit Circumpolar Council definition of, 7
　outsiders' use without compensation, 71–72
　programs documenting, 56
　role in CBM, 16, 26
　value of, x
Indigenous Observation Network (ION) [Yukon River], 23*t*, 42, 49–51, 88*t*
Indigenous peoples of Arctic. *See also* communities
　and Arctic Science Ministerial meeting of 2016, 28
　and CBM recommendations, 3*f*
　impact of climate change on, 1
　partnering with government agencies, 42
　risk of marginalization in sharing of data, 45
Indigenous rights, protection of, xi. *See also* intellectual property rights
　awareness of protocols for, 71
　challenges in, 70–72, 76, 84
　and communities' cooperation, 70
　and free, prior and informed consent (FPIC), xi, 45, 53, 57, 58, 71, 76, 84
　good practice in, 53–57, 58
　increasing demands for, 70–71
　interventions to improve, 72–73, 76
　knowledge and information rights, 53–54, 57
　land and resource-related rights, 53
　as motive for CBM participation, 53, 55*f*, 71

and right to shape decision-making on natural resources, 56*f*
Integrated Arctic Observation System Project (INTAROS), 2, 9
intellectual property rights of Indigenous people, 21, 45, 53–54, 58, 68
　reported violations of, 70–72
　scientists' lack of knowledge about, 76
International Centre for Reindeer Husbandry (Norway), 40
International Convention for the Regulation of Whaling (ICRW), 23*t*
International Council for the Exploration of the Sea (ICES), 79
international environmental agreements
　Arctic CBM contribution to meeting goals of, x, 9, 20, 22, 23*t*, 26
　countries' action in response to, 21–22
　monitoring effectiveness of, 22
internet access for CBM programs, 83
Inuit Circumpolar Council, definition of Indigenous knowledge, 7
Inuvialuit Settlement Region Community-Based Monitoring Program (CBMP) [Canada], 42–43, 51–52
ION. *See* Indigenous Observation Network

J

Johnson, Noor. *See Community-based Monitoring and Indigenous Knowledge in a Changing Arctic* (Johnson et al.)

K

Kitikmeot Heritage Society, 34

L

Local Environmental Observer (LEO), 23*t*, 88*t*
local knowledge. *See* Indigenous and local knowledge
local stakeholder inclusion, increasing emphasis on, 77–78

M

management agencies of Arctic CBM programs
　adversarial relations with local communities, 61
　communication with participants, importance of, 32
　and effort needed for collecting data, importance of understanding., 32

management agencies of Arctic CBM programs *(continued)*
   and Indigenous and local knowledge, insufficient incorporation of, 79
   lack of interest in community input, xi, 60–63, 75
      causes and effects of, 60–62, 75
      interventions for, 62–63
   most-important, 79
   planning of, 29–32
   staff turnover
      causes of, 74
      problems created by, 36, 38, 57
   training, need for improvement in, 62–63, 80–81
   work with CBM programs, need for greater expertise in, 79
Marion Watershed, 23*t*, 88*t*
muskox in Greenland, decentralized management of, 3*f*, 79–80

N

NAFO. *See* Northwest Atlantic Fisheries Organization
NAMMCO. *See* North Atlantic Marine Mammal Commission
National Oceanic and Atmospheric Administration, Arctic Report Card (2019), 82–83
National Weather Service, 41
Nenets Autonomous Okrug, 33*f*
Nordland Ærfugl, 12, 23*t*, 88*t*
North Atlantic Marine Mammal Commission (NAMMCO), 79
Northwest Atlantic Fisheries Organization (NAFO), 79
Nunaliit software, 34
Nunavut Wildlife Management Board, Community-Based Monitoring Network, 37

O

Obama, Barack, 28
Oral History, 23*t*, 88*t*
organizational and support structures
   challenges in, 73–75, 76
      causes and effects of, 73–74
      interventions to resolve, 74–75, 76
   importance of, 84
OSPAR. *See* Convention for the Protection of the Marine Environment of the North-East Atlantic

P

Paris Agreement, and value of CBM data, 83
participants in Arctic CBM
   age of, 12, 24
   compensation for, 12, 36–37
   gender of, 12, 24
   management communication with, importance of, 32
   motives for participation, 11–12, 24, 53, 55*f*
   number per program, 11–12, 24
   overwork of, 32, 66, 75
   training for, 47*f*
participants' engagement
   lack of, causes and effects, 66–67, 75
   need for research on, 83–84, 85
   in programs not tied to community priorities, 64–65, 66
   strategies for increasing, 33, 35–38, 67–68, 75
   vital importance of, 83
Pilot Whale, 23*t*, 88*t*
Piniakkanik Sumiiffinni Nalunaarsuineq (PISUNA) [Greenland]
   area covered by, 14
   awards won by, 28
   contact information, 89*t*
   high quality of data, 49, 50*t*
   independence of, 42
   and Indigenous rights, 53
   and international environmental agreements, 23*t*
   management impact of, 39
Piniarneq, 23*t*, 88*t*
PISUNA. *See* Piniakkanik Sumiiffinni Nalunaarsuineq
POP. *See* Convention on Persistent Organic Pollutants

Q

questionnaire, 91–99
   adaptation of terminology to local use, 9
   number of questions, 9
   response rate, 4
   types of questions, 4

R

Ramsar Convention on Wetlands of International Importance, 23*t*
remote locations, need for increased monitoring in, 81–82
Renbruksplan, 23*t*, 89*t*

resource management. *See also* international environmental agreements
decentralized
existing programs, 79–80
as goal, 79
impact of CBM programs on, x, xi, 19–20, 25, 26
Indigenous peoples' right to shape decision-making on, 56*f*

S

SAON. *See* Sustaining Arctic Observing Networks
scientific bodies, international, limited focus on Indigenous knowledge, 61–62
scientist-executed monitoring programs
vs. community-based monitoring
asymmetric power issues and, 45
as complementary, x, 25, 43
and data format incompatibility, 68–70, 72, 76, 83
differences in method, 8*b*
equivalent validity of data, xi, 46, 49–51, 50*t*
coordinating with
good practice in, 41–42, 45
mutual trust as necessary to, 46
scientists, need for information on CBM, 78–79
scientists working with CBM programs
incentive system for, 61, 63, 73
lack of knowledge about larger context of work, 61
measures to assure data accuracy, 48, 58
SDGs. *See* United Nations Sustainable Development Goals
Sea Ice for Walrus Outlook (SIWO), 23*t*, 40–41, 89*t*
Seal Monitoring, 23*t*, 89*t*
Seasonal Ice Zone Observing Network, 54
SIWO. *See* Sea Ice for Walrus Outlook
SIZONet, 34
Snowchange Cooperative (Finland), 34, 39–40, 53
socially-sustainable development, use of CBM to promote, 80
social media, and participant engagement, 33. *See also* Facebook
Summer Farmers and Small Herders, 23*t*, 89*t*
support for Arctic CBM
funding sources, 20, 21, 25
scientific support, 20
Sustaining Arctic Observing Networks (SAON), ix
studies on, 2

sustaining of CBM programs
challenges in, 21, 84
good practice in, 36–38, 57
planning for, 31, 74
Svalbard Social Science Initiative (Svalbard), 74

T

Thcho Government (Canada), 53
Tkarihwaie:ri Code, 71

U

UArctic Thematic Network on Collaborative Resource Management, 62–63
United Nations Framework Convention on Climate Change (UNFCCC).
Arctic CBM contribution to meeting goals of, 22, 23*t*
and value of CBM data, 83
United Nations Sustainable Development Goals (SDGs)
Arctic CBM contribution to meeting, x, 5, 22–24, 25*t*, 48
growing emphasis on local stakeholder inclusion and, 77
University of Alaska Fairbanks
and Arctic and Earth SIGNs program, 54–56
Winterberry Project, 37*f*
University of the Faroe Islands, data storage by, 34
Ureu-Wau-Wau tribe, 72
US Geological Survey (USGS), 42, 44*t*, 49

W

Walrus Haulout Monitoring, 23*t*, 89*t*
web-based knowledge management platforms
advantages and disadvantages of, 36
as means of publicizing value of CBM data, 83
Nunaliit software for, 34
potential of, as only partially realized, 79
SIZONet, 34
use by CBM programs, x, xi, 34–36
websites of Arctic CBM programs, data on, 21, 87*t*–89*t*
Western Arctic Beluga Health Monitoring, 42–43
Wildlife Triangles, 23*t*, 89*t*
Wild North, 23*t*, 89*t*
WinterBerry, 23*t*, 89*t*
workshops with CBM participants
data sharing, participants' emphasis on value of, 44–45

workshops with CBM participants (*continued*)
    funding for, 2
    good practices gleaned from, 27
    issues discussed in, ix
    locations of, ix, 2, 5, 6*t*, 27
    as means of publicizing CBM benefits, 82
    participants, 5–6
    statement on value of CBM, 1–2
    topics discussed in, 3*f*, 6

Y

Yukon River Inter-Tribal Watershed Council (YRITWC), 49–51, 53
Yukon River Water Quality Plan, 42
Yup'ik Environmental Knowledge Project and Atlas, 34

CPSIA information can be obtained
at www.ICGtesting.com
Printed in the USA
LVHW071740230222
R17181800001B/R171818PG711424LVX00001B/1